The SCALE of the UNIVERSE

The SCALE of the

Jeffrey Bennett

Big Kid Science
Boulder, Colorado

Published by
Big Kid Science
Boulder, CO
www.BigKidScience.com
Education, Perspective, and Inspiration for People of All Ages

Editing: Lifland et al., Bookmakers
Composition and design: Side By Side Studios

ISBN: 978-1-937548-94-0
(E-book ISBN: 978-1-937548-95-7)

Distributed by IPG
Order online at www.ipgbook.com
or toll-free at 800-888-4741

Cover images: NASA, ESA, CSA, STScI (James Webb Space Telescope); Michael Carroll; NASA (Apollo 17)

Also by Jeffrey Bennett

For Children

Max Goes to the Moon
Max Goes to Mars
Max Goes to Jupiter
Max Goes to the Space Station
The Wizard Who Saved the World
I, Humanity
Totality! An Eclipse Guide in Rhyme and Science

For Educators and the Public

Beyond UFOs: The Search for Extraterrestrial Life and Its Astonishing Implications for Our Future
Math for Life: Crucial Ideas You Didn't Learn in School
What Is Relativity? An Intuitive Introduction to Einstein's Ideas, and Why They Matter
On Teaching Science: Principles and Strategies That Every Educator Should Know
A Global Warming Primer: Pathway to a Post-Global Warming Future

Middle/High School Textbooks

Earth and Space Science (online at grade8science.com)

High School/College Textbooks

The Cosmic Perspective
The Essential Cosmic Perspective
The Cosmic Perspective Fundamentals
Life in the Universe
Using and Understanding Mathematics: A Quantitative Reasoning Approach
Statistical Reasoning for Everyday Life

To Lisa, Grant, and Brooke: You inspire me with the example you set of how to treat all of our fellow humans with empathy, kindness, and respect. I could not have been more fortunate in having you as my family.

Brief Contents

Detailed Contents

Preface

Have you ever wondered how big the universe really is or how we fit into it? These questions have been asked throughout history by people of virtually every race, religion, culture, and nationality. Indeed, they seem integral to the human experience, which probably explains why children everywhere love to learn about our planet and space. This book is designed to help you understand the evidence-based answers that modern science now gives us to these ancient questions, along with the implications that these answers may hold for us as a species.

I hope that this book will be of interest to readers of all ages and backgrounds, but I've tried to organize it in a way that will make it of particular interest to students, parents, and teachers:

Students: This book should be accessible to anyone in about middle school and up.

Parents: This book will prepare you to answer many of the questions about our universe that your kids will almost certainly ask you as they grow up.

Elementary School Teachers: This book will help you to answer the questions you will inevitably get from students, no matter what grade level you teach.

Middle and High School Teachers: For those teaching science, the more detailed discussions in this book, including the Teacher Notes at the end, should help you understand the material you are expected to teach as part of general science standards. For those teaching other subjects, you will find numerous ways in which you can integrate the topics of this book into your teaching, which will likely help you keep students engaged with your subject matter.

Everyone Else: The question-and-answer format of the book, along with its set of endnotes (Teacher Notes), will allow you to choose the depth to which you'd like to go in learning the answers to your questions about the scale of the universe.

With regard to the origins of this book, I have been teaching and writing about astronomy throughout my career, and the scale of the universe has always been my favorite topic. It never ceases to boggle the mind, even for someone like me who has spent much of a lifetime thinking about it. I started developing methods for explaining scale during my earliest teaching days, which led to my work on the Colorado Scale Model Solar System on the University of Colorado Boulder campus and then on the Voyage scale model solar system project (see Figure 3.8). The scale of the universe is also the topic that opens my college-level astronomy textbook (*The Cosmic Perspective*), and some aspects of scale are discussed in many of my other books. Until now, however, I had never put all of these ideas about scale together in one place, but the importance of the topic made me decide it would be worth the effort to do so. I hope you will find that the effort has been successful.

Finally, I'll note that while scale is my favorite astronomical topic, there are many others that are also quite interesting and that I believe it would be useful for everyone to know. I therefore hope that this book will be only the first in a series (the "Big Kid Science Knowledge Series") of similarly formatted books on topics in Earth science, space science, and astronomy. For example, future topics might include the cause of the seasons, moon phases and eclipses, Earth system science, black holes, and more. If you have thoughts about the value of such a series — or any comments on this book — please feel free to reach out to me through the email address that you'll find on my websites (**jeffreybennett.com** or **bigkidscience.com**).

Jeff Bennett
Summer 2025

Using This Book

Please note the following key elements of this book's structure:

- Each chapter begins with a short introduction and then proceeds in question-and-answer format.
- The Q&A has two levels of questions and answers:
  - The first level, set in this normal font, is for the key ideas needed to understand the scale of the universe.
  - The second level, set in this smaller font, is for more detailed discussion that you can treat as optional, depending on the depth to which you'd like to go.
- Endnotes, designated "**TN**" (for Teacher Notes), are aimed especially at classroom teachers but may also be of interest to others.
- **Building a Cosmic Perspective** features are designed to help you think about how scale ideas might affect your perspective on yourself, our species, and our planet. Note: Teachers using inquiry-based learning methods may wish to use these for short writing assignments.

Figure 1.1. This incredible image from the James Webb Space Telescope (JWST) shows thousands of galaxies in a region of the sky so small that you could cover it with a grain of sand held at arm's length. Note: The JWST records infrared light, which our eyes cannot see, so the colors in this image are artificial but chosen to mimic what we might see if the light were visible to us.[**TN1**] Credit: NASA, ESA, CSA, STScI.

1

Introduction

Picturing Space and Time

Look carefully at the remarkable image in Figure 1.1. This photo, taken by the James Webb Space Telescope (JWST), shows a piece of the sky so small that you could block your view of it with a grain of sand held at arm's length. Yet it encompasses an almost unimaginable expanse of both space and time. Nearly every object within it is an entire *galaxy* filled with billions of stars. (The exceptions are the objects with spikes, which are foreground stars in our own Milky Way Galaxy.[TN2]) Moreover, based on discoveries of the past several decades, we can say with great confidence that most of the stars in these galaxies have orbiting planets, and it is likely that many of these planets are similar in nature to Earth.

Even if you don't yet know much about the scale of a solar system or galaxy, Figure 1.1 probably already gives you some sense of the vast size of our universe. But I've said that this image also shows a vast expanse of *time*. The reason is that different galaxies are at different distances, which means it has taken different lengths of time for their light to reach us. The most distant galaxies in the image (appearing only as small red smudges) are more than 13 billion light-years away, meaning that their light has taken more than 13 billion years to reach us.[TN3] We therefore conclude that Figure 1.1 is showing us objects as they appeared over a span of more than 13 billion years of time — a time span that encompasses most of the history of the universe.

The primary goal of this book is to help you make sense of the scale of space and time. It would be easy to simply state the numbers — for example, it is easy to say "13 billion years" — but it is much more challenging to wrap your mind around what such numbers really mean. After all, we are talking about distances and times that are *literally* astronomical. We'll therefore take a methodical approach, starting with a review of our cosmic address in Chapter 2, moving on to the scale of space in Chapter 3 and the scale of time in Chapter 4, and concluding with a brief summary in Chapter 5.

Along the way, we'll see many astonishing implications of the scales we discuss. For example, we'll see how the sheer number of planets forces us to think about possibilities for life beyond Earth, while the vast distances between stars provide humbling lessons about the technology that would be required for interstellar travel. But I especially hope that this book will help you think about what I believe are the two most important lessons we can learn from the scale of the universe:

1. We are *not* the center of the universe. This is a lesson that can be difficult to absorb, but I believe that if everyone understood it, we would all recognize that no individual is any more central or important than any other, and we would therefore all treat each other with greater empathy, kindness, and respect.
2. We are responsible for the future of our planet Earth. If we want our modern civilization to endure, we must learn to take good care of our small and fragile home in space.

Building a Cosmic Perspective Take another close look at Figure 1.1. In what sense does this image offer a "cosmic perspective" that can change the way we think about ourselves and our planet? Continue to think about this image and its meaning as you proceed through this book.

What exactly *is* a light-year?

A light-year (ly) is the *distance* that light can travel in 1 year, which is about 10 trillion kilometers (6 trillion miles). You can easily confirm

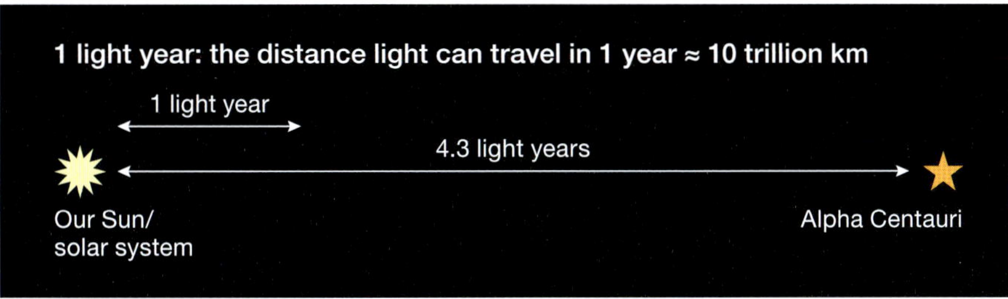

Figure 1.2. This figure summarizes the meaning of a light-year, showing the 4.3-light-year distance to Alpha Centauri, the nearest star besides the Sun (more technically, to the brightest of the three stars found in the Alpha Centauri system).

this fact for yourself simply by multiplying the speed of light, which is about 300,000 kilometers per second (186,000 miles per second), by the number of seconds in 1 year.[TN4]

Note that light travels extremely fast by earthly standards. If you could make light go in circles, it could circle Earth nearly eight times in a single second! Nevertheless, even light takes significant time to travel the vast distances in space. Light takes a little more than 1 second to reach Earth from the Moon and about 8 minutes to reach Earth from the Sun. Stars are so far away that their light takes years to reach us, and the light of distant galaxies can take millions or even billions of years to reach us. That is why we generally use light-years to describe the distances of stars and galaxies.[TN5]

The light-year is a convenient unit for astronomy not only because it represents a very large distance but also because knowing an object's distance in light-years tells us how long it takes for light from the object to reach us. For example, when we say that the nearest star (besides the Sun) is about 4.3 light-years away (**Figure 1.2**), it also means that light from this star takes about 4.3 years to reach us.

One note of caution: The relationship between the *distance* represented by a light-year and the time it takes for light to travel that distance sometimes leads people to misuse "light-years" to refer to times rather than distances. If you are unsure whether the term *light-year* is being used correctly in a particular context, try testing the statement by using the fact that 1 light-year is about 10 trillion kilometers. For example, imagine that you heard a student say "It will take me light-years to finish this homework!" If you substitute its equivalent distance for "light-years," the statement becomes "It will take me 10 trillion kilometers to finish this homework," which clearly does not make sense.

Are there other "light units," such as light-seconds or light-minutes?

Yes, although they are less commonly used. They are all defined in the same basic way as light-years. For example, a light-second is the distance that light can travel in 1 second, which is about 300,000 kilometers (186,000 miles); we can therefore say that the Moon's average distance of 384,000 kilometers (239,000 miles) is about 1.3 light-seconds. Similarly, a light-minute is the distance that light can travel in 1 minute (the Sun is about 8 light-minutes away), a light-day is the distance that light can travel in 1 day, and so on.

How does light travel time allow us to see into the past?

As we've already discussed, images like those in Figure 1.1 allow us to see objects as they were at various times in the past. This is always the case to some extent, as it is a simple consequence of the fact that it takes time for light to travel vast cosmic distances. For example, the fact that light from the Moon takes a little more than 1 second to reach Earth means that we're always actually seeing the Moon as it appeared a little over 1 second ago. Similarly, because light takes about 8 minutes to travel from the Sun to Earth, we are always seeing the Sun as it looked about 8 minutes ago. If we observe a solar flare appear suddenly on the Sun, it actually occurred about 8 minutes earlier.

The effect is much more dramatic for stars and galaxies. Consider Sirius, the brightest star in the night sky, which is about 8 light-years away. This means it takes light from Sirius about 8 years to reach us, so we see Sirius as it was about 8 years ago. If we look at a star that is 100 light-years away, we see it as it was 100 years ago. If we look at a galaxy that is 10 million light-years away, we see it as it was 10 million years ago. And so on. The general rule is that *the farther away we look in space, the further back we look in time.*[TN6]

What units do astronomers use to describe distances in our solar system?

The most distant planets (and dwarf planets) in our own solar system are only a few light-hours away, which means units of light-years are too big to be convenient. Astronomers therefore usually measure distances within our solar system by comparing them to the average Earth-Sun distance, which is about 150 million kilometers (93 million miles) and is called an *astronomical unit,* abbreviated as AU. For example, the distance from the

Sun to Neptune is about 30 times as far as the distance from the Sun to Earth, so we say that Neptune is located about 30 AU from the Sun. We won't make much use of AU in this book, but you'll encounter this unit if you do further study of astronomy.

Why are astronomers confident that most stars have planets?

In discussing Figure 1.1, I stated that "we can say with great confidence that most of the stars in these galaxies have orbiting planets." You might wonder where this confidence comes from. The answer is statistics based on discoveries of planets orbiting other stars, often called *exoplanets* because the prefix *exo* means "outside" (or "external") and these planets lie outside of our own solar system.

Exoplanets are extremely difficult to detect, primarily because (as we'll discuss in Chapter 3) all other stars are incredibly far away. In addition, even in cases in which exoplanets are bright enough for our telescopes to detect in principle, their light is usually drowned out by the far brighter light of the stars they orbit. Nevertheless, astronomers have already identified more than 5,000 exoplanets, and this is enough to allow statistical estimates of what we would find if we were able to search for planets around all stars. Current data suggest that at least about 70% of all stars have planets, and the actual percentage may well be 90% or even higher. We therefore conclude that most — and possibly almost all — stars have planets.[TN7]

Perhaps the more interesting question is how many planets might be similar in nature to Earth, meaning that they could at least potentially have life. This question is more difficult, because we have very limited information about the nature of the exoplanets discovered to date. Nevertheless, we know the approximate sizes and orbital distances (around their stars) of most of these planets, so we can again use statistics. The results suggest that, at least on average, most star systems are likely to have at least one planet that is similar in size to Earth and is orbiting its star at a distance at which it might have an Earth-like surface temperature. As we'll discuss in more detail later, this means that while we don't yet know of life anywhere in the universe besides Earth, the sheer number of potential homes to life makes for astonishing possibilities.

Building a Cosmic Perspective As recently as the early 1990s, we did not know for sure whether there were any planets outside our own solar system. Now we know that most stars have planets. Should this change in human knowledge affect the way we look at the stars we see in the night sky? Why or why not?

How do astronomers detect exoplanets?

Given the enormous difficulty of detecting planets orbiting other stars, you might wonder how astronomers manage it at all. In fact, there have been only a handful of cases to date in which astronomers have managed to get actual telescopic images of exoplanets, and these have had such low resolution (usually only a pixel or two) that they don't offer any real detail aside from telling us that the planets exist. Instead, nearly all known exoplanets have been detected *indirectly* by noticing the effect that these planets have on their stars rather than by seeing the planets themselves.

There are two major approaches to the indirect detection of planets: (1) observing subtle motions of a star that are caused by the gravitational effects of orbiting planets and (2) observing temporary dips in a star's brightness that occur when one of its planets passes in front of the star as viewed from Earth. The vast majority of known exoplanets (as of 2025) have been discovered with the second approach.

Figure 1.3 shows how this second approach works. It is called the *transit method*, because the word *transit* means "to pass across." Keep in mind that stars are much too far away for us to actually observe the planet moving across the face of the star. Instead, we learn that the planet exists by noticing how the star's brightness dips as the planet transits in front of it. A transit event typically lasts a few hours.

You might wonder how astronomers can know that a dip in brightness is due to a planetary transit rather than something else, such as an intrinsic change in the star itself. The answer is that transits repeat with every orbit. For example, if a planet takes 30 days to orbit its star, then we would see the transit repeat every 30 days. No other known phenomenon could cause a star's brightness to change with such regularity for just a few hours at a time.

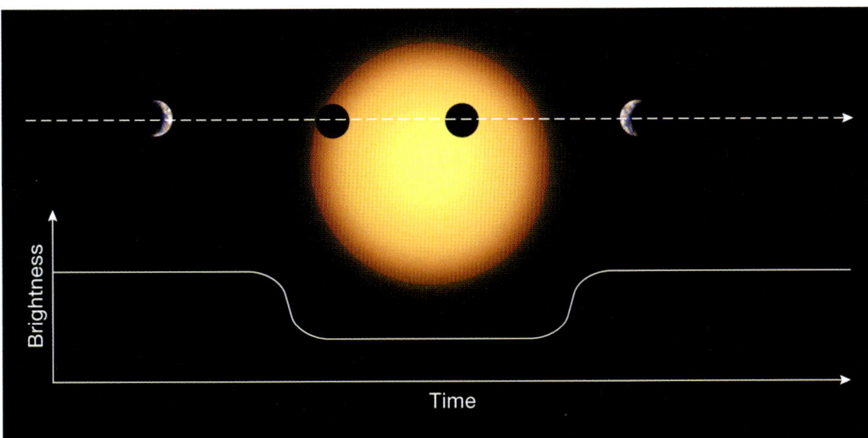

Figure 1.3. This diagram shows the basic idea of the transit method. If a planet happens to have an orbit aligned so that it transits in front of its star as viewed from Earth (top), a graph of the star's brightness (bottom) will show a dip during the transit (which typically takes a few hours). Credit: Michael Carroll.

The biggest limitation of the transit method is that it will work only if the planet happens to have an orbit that is aligned "just right" for the planet to pass directly in front of the star as seen from Earth. This type of chance alignment is fairly rare, occurring in only about 1% of all star systems. As a result, finding a few thousand exoplanets with the transit method has required careful study of hundreds of thousands of star systems. On the plus side, these large studies provide a statistical estimate of how common planets are. We simply compare the percentage of stars expected to have the necessary alignment for transits to the percentage of stars for which we observe transiting planets.

We can learn a surprising amount about exoplanets with indirect methods. The transit method tells us the planet's orbital period (it is the time from one transit to the next), from which it is possible to calculate its average orbital distance from its star. The transit method also allows us to estimate the planet's diameter from the amount of dimming it causes in its star (the larger the diameter, the more light it blocks). If we then follow up a transit discovery with observations of how the planet's gravity affects its star, we can learn the planet's mass, density, and more. This is how astronomers have learned about enough planets to make statistical estimates of how many might be "Earth-ish," meaning similar enough in size and orbit to Earth to make us think that they *might* be Earth-like in the sense of having oceans and an atmosphere that could potentially support life.

Figure 2.1. This illustration shows the four major levels of structure that make up our "cosmic address." The solar system box uses zoom-outs for the planets because they would be too small to see on the scale used for their orbits. Credit: Adapted from *The Cosmic Perspective*.

2

Our Cosmic Address

The first step in learning the scale of the universe is to become aware of the basic levels of structure of the universe. Up until only about 400 years ago, most people believed that we were located at the center of the universe and that the Sun, Moon, planets, and stars all circled around us once each day. Today, we know that Earth is just one planet orbiting our Sun, that our Sun is just one star in our Milky Way Galaxy, and that our galaxy is just one of billions of galaxies in the universe (**Figure 2.1**).

A good way to remember these ideas is to imagine that, to take a quote from *Star Wars*, you were magically transported to a galaxy "far, far away." If you wanted to send a postcard to a friend at home, you'd need to address it something like this:

Name
Street Address
City/State/Country
Earth
The Solar System
The Milky Way Galaxy
The Universe

The last four lines above make up what we might call our *cosmic address* (**Figure 2.2**), because they describe the location of every human being who has ever lived:

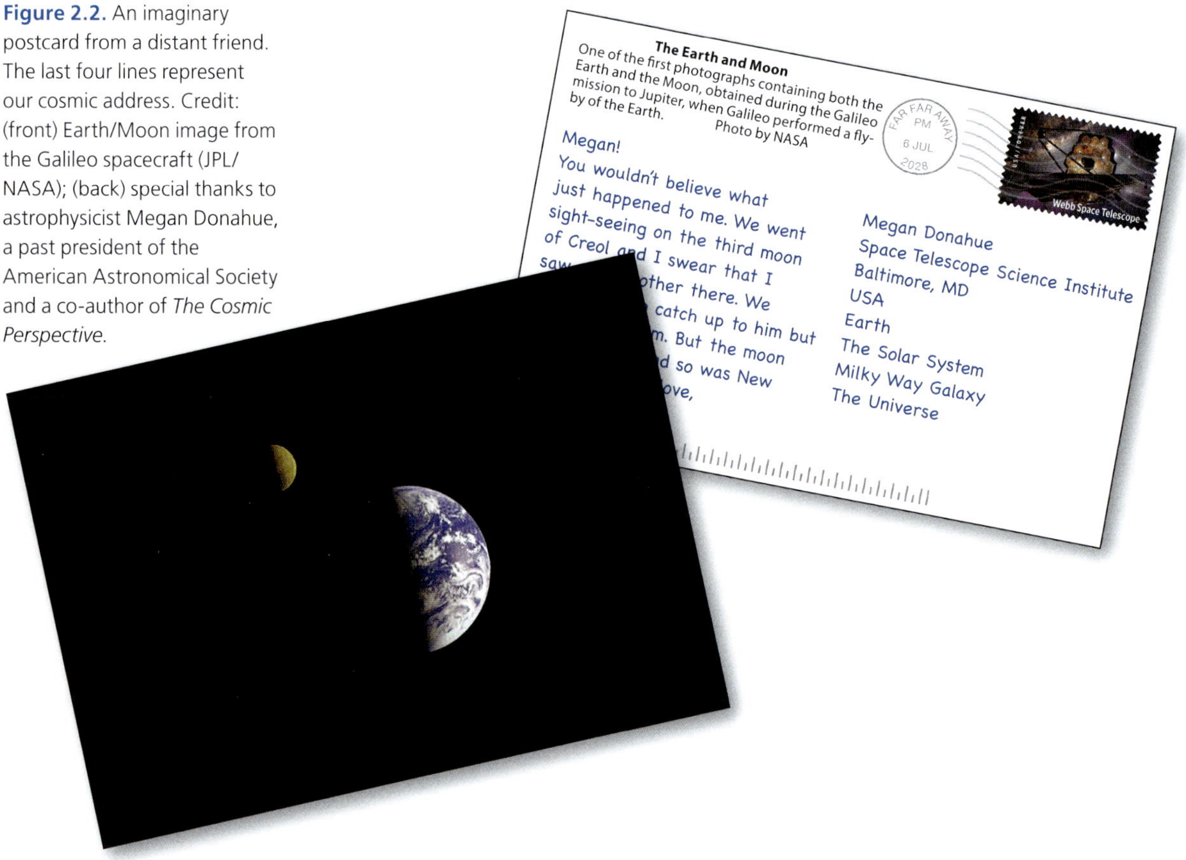

Figure 2.2. An imaginary postcard from a distant friend. The last four lines represent our cosmic address. Credit: (front) Earth/Moon image from the Galileo spacecraft (JPL/NASA); (back) special thanks to astrophysicist Megan Donahue, a past president of the American Astronomical Society and a co-author of *The Cosmic Perspective*.

- We live on a *planet* called Earth.
- Earth is part of our *solar system*, which consists of the Sun and the family of planets, moons, and other objects that orbit it.
- Our solar system is one of at least 100 billion star systems that make up our Milky Way Galaxy.
- Our Milky Way Galaxy is one of at least 100 billion galaxies in our universe.

We'll briefly discuss each of these major levels of structure in the rest of this chapter.

Building a Cosmic Perspective What's *your* full address, including your cosmic address?

What is a planet?

The first line of our cosmic address is that of our *planet* called Earth, which we often refer to as our "home planet." But what exactly do we mean by the word *planet*?

Astronomers argue quite a bit about a precise definition, but for most purposes we can think of a planet as an object that orbits a star (such as our Sun) and is at least moderately large (currently taken to mean at least about a third the diameter of Earth). With this definition, astronomers today identify eight planets in our solar system: Mercury, Venus, Earth, Mars, Jupiter, Saturn, Uranus, and Neptune (**Figure 2.3**).[TN8] As mentioned earlier, astronomers have also identified more than 5,000 planets orbiting other stars, which means we now know of far more planets outside of our own solar system than within it.

Why don't we count Pluto as a planet anymore?

You're probably aware that we used to say there were *nine* planets in our solar system. The ninth was Pluto, which was demoted to the status of "dwarf planet" in 2006.

To understand what happened, you have to go back to Pluto's discovery in 1930. Scientists initially assumed that Pluto was large, perhaps like Uranus or Neptune, so it seemed natural to consider Pluto as the ninth planet. Even then, however, Pluto seemed a bit of a misfit among the planets, because its orbit around the Sun is both much more elliptical (stretched out) and more inclined to the plane of Earth's orbit (the ecliptic

Figure 2.3. The eight official planets in our solar system, shown in order of distance from the Sun. The planet sizes (but not their distances) are to scale. Credit: NASA/Lunar and Planetary Institute.

Figure 2.4. Pluto (far right) compared in size to Earth and the Moon. Pluto's smallness stands out even more in terms of its mass, which is 450 times smaller than Earth's and 25 times smaller than Mercury's (the smallest planet if we don't count Pluto). Credit: Earth from NASA/Apollo 17, Moon from Gregory H. Revera (used under Creative Commons license), Pluto from NASA/New Horizons.

plane) than that of any other planet. Over the next few decades, scientists gradually came to realize that Pluto was much smaller than originally thought, a fact confirmed when the 1978 discovery of its moon Charon allowed precise determination of its mass and size (**Figure 2.4**).[TN9] Most significantly, scientists then began to discover many "Pluto-ish" objects orbiting the Sun in the same region of the outer solar system, raising the question of whether any of these objects might also be considered planets.

The situation came to a head in 2005 with the discovery of an object called Eris, which is about the same size as (and about 27% larger in mass than) Pluto. At this point, astronomers realized they had to make a decision: Either Eris and other similar-size objects would need to be added to the planet list or Pluto would need to be kicked off. Although the decision remains controversial, Pluto and Eris were both put into a new category called *dwarf planets*, which essentially refers to objects that are round like planets but too small to make the full planet list. A total of five dwarf planets have been identified to date (as of 2025). Besides Pluto and Eris, these are the asteroid Ceres and two other outer solar system objects known as Haumea and Makemake. However, it's likely that the full list of dwarf planets in our solar system will ultimately be much larger.[TN10]

Perhaps the most important take-away from all this is that Pluto and other worlds remain the same no matter what we call them. In this sense, the entire controversy over the definition of a planet is simply a reminder of the difference between the fuzzy boundaries of nature and the human preference for categories. After all, the argument about whether Pluto should be called a planet, dwarf planet, or something else is not so different from the argument you may hear about whether a particular waterway should be called a creek, stream, or river.

Building a Cosmic Perspective Even without Pluto, the planets of our solar system span a wide range of sizes. For example, the storm on Jupiter known as the Great Red Spot (visible near Jupiter's lower right in Figure 2.3) could by itself swallow up the entire Earth. If there were intelligent beings on Jupiter, do you think they would consider Earth to be large enough to count as a planet?

Did the definition of *planet* ever change before?

Yes. The word *planet* comes from an ancient Greek term meaning "wanderer," and it originally referred only to objects that appear to "wander" relative to the constellations in the sky. Because we do not see Earth in our own sky, Earth was not considered a planet in ancient times. Earth therefore became a planet only when the discovery that Earth orbits the Sun forced a redefinition of the word.[TN11]

By the way, did you know that the ancient definition of *planet* explains why we have seven days in a week? Ancient people counted seven "planets" that they observed to wander among the constellations: the five planets easily visible to our naked eyes (Mercury, Venus, Mars, Jupiter, and Saturn) plus the Sun and the Moon. These seven objects give us our week, with each "planet" having its own day:

- Sunday, Monday ("Moon day"), and Saturday ("Saturn day") are fairly obvious in English.
- The other four days are less obvious in English, but easy to see in many other languages. For example, if you know Spanish, you will recognize that
  - Tuesday is martes, which makes it "Mars day."
  - Wednesday is miércoles, which makes it "Mercury day."

- Thursday is jueves, which makes it "Jupiter day." (Fans of Marvel movies will also recognize Thursday as "Thor's day," and Thor is the Norse equivalent of the Roman god Jupiter.)
- Friday is viernes, making it "Venus day."

Building a Cosmic Perspective The planet Uranus is faintly visible to the naked eye, but it is so dim and moves so slowly through the constellations (because it is so far from the Sun) that it was not recognized as a planet in ancient times. But suppose that Uranus had been recognized as a planet. Do you think we would have eight days a week instead of seven?

Why are planets round?

You can see in photos (such as those in Figure 2.3) that Earth and all the other planets are round.[TN12] Indeed, as noted earlier, roundness is part of the definition of a planet. But *why* are planets round? The answer is *gravity.*

The general reason is easy to understand: Gravity always pulls toward a center, and forces that pull toward a center tend to make objects spherical (**Figure 2.5**). Indeed, the bigger question may be why smaller objects, such as most asteroids and comets, often have other shapes.

The differing shapes arise because the strength of gravity depends on an object's size and mass. Small, low-mass objects have weak gravity and can therefore retain whatever shape they happen to take on as they form (**Figure 2.6a**). More massive objects generally have stronger gravity, and when gravity is strong enough it can gradually reshape even solid rock into a sphere. That is why all large objects — including stars, planets, dwarf planets, and large moons — are round (**Figure 2.6b**).[TN13]

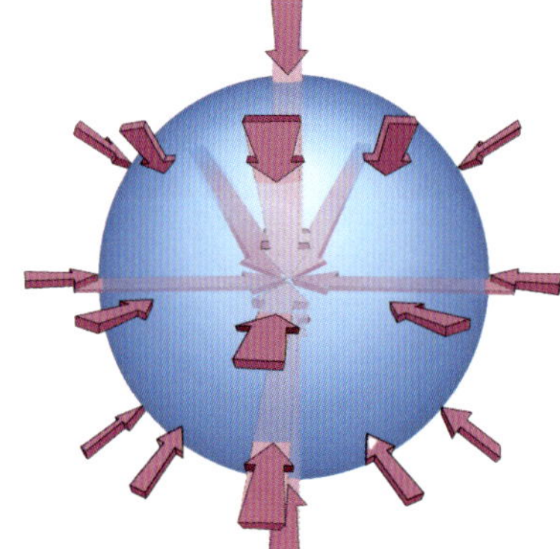

Figure 2.5. A force that pulls toward a center, with equal strength in all directions, will tend to make an object spherical in shape. Credit: Xana Sound.

(a) This montage show several small asteroids that have been visited by spacecraft. Notice that, somewhat like potatoes, they can have almost any shape.

(b) This montage shows selected objects (ranging in size from dwarf planets like Ceres and Pluto to stars like the Sun) with gravity strong enough to make them round.

Figure 2.6. The weak gravity of small objects allows them to have almost any shape, while the strong gravity of larger objects makes them spherical. Credit: All photos from NASA/JPL/ESA.

What is a solar system?

The second line in our cosmic address is our *solar system*, which consists of the Sun and all the objects that orbit it.[TN14] In other words, our solar system is essentially the Sun's family, and the family members include

- our star, the Sun
- the eight planets: Mercury, Venus, Earth, Mars, Jupiter, Saturn, Uranus, and Neptune
- more than 400 known *moons*, each of which orbits one of the eight planets

- the dwarf planets (including Pluto) and vast numbers (billions to trillions) of smaller objects that orbit the Sun independently. We refer to these smaller objects generically as "small solar system bodies" but often further categorize them as either *asteroids* or *comets*.
  - Asteroids are small bodies made mainly of rocky material; most asteroids orbit within the *asteroid belt* located between Mars and Jupiter.
  - Comets are small bodies that contain lots of ice[TN15] in addition to rock; most comets orbit either in the *Kuiper belt* (*Kuiper* rhymes with *piper*) beyond Neptune or in the much more distant *Oort cloud* (*Oort* rhymes with court).

Other stars have their own families of planets and other orbiting objects, which means the stars we see in the night sky and in photos are really other solar systems (also called "star systems" or "planetary systems").

Figure 2.7 shows the general layout of our solar system, including the donut-shaped regions of the asteroid belt and Kuiper belt. The more spherically shaped Oort cloud is too far away to be shown on this scale. Notice that all the planets orbit the Sun in the same direction and in approximately the same plane. The same is also true of most objects in the asteroid and Kuiper belts. This orbital pattern is not an accident. Instead, as we'll discuss shortly, it is a direct result of the way in which our solar system formed. First, however, let's start with three simpler questions to be sure you understand the definitions of the objects listed above (except *planet*, which we've already discussed).

What is the Sun?

The Sun is a star. It looks different (larger and far brighter) than the stars we see at night only because it is so much closer to us. Like all stars, the Sun is essentially a giant ball of extremely hot and dense gas,[TN16] held together by its own gravity. The Sun is by far the most massive object in our solar system, outweighing all the planets combined by a factor of nearly 1,000 (which means it contains more than 99.8% of the solar system's total mass). That is why its gravity is strong enough to hold the planets and other objects in orbit around it.

In case you are wondering why the Sun shines, it is because its surface is so hot, and hot objects always give off light; that is also why flames and

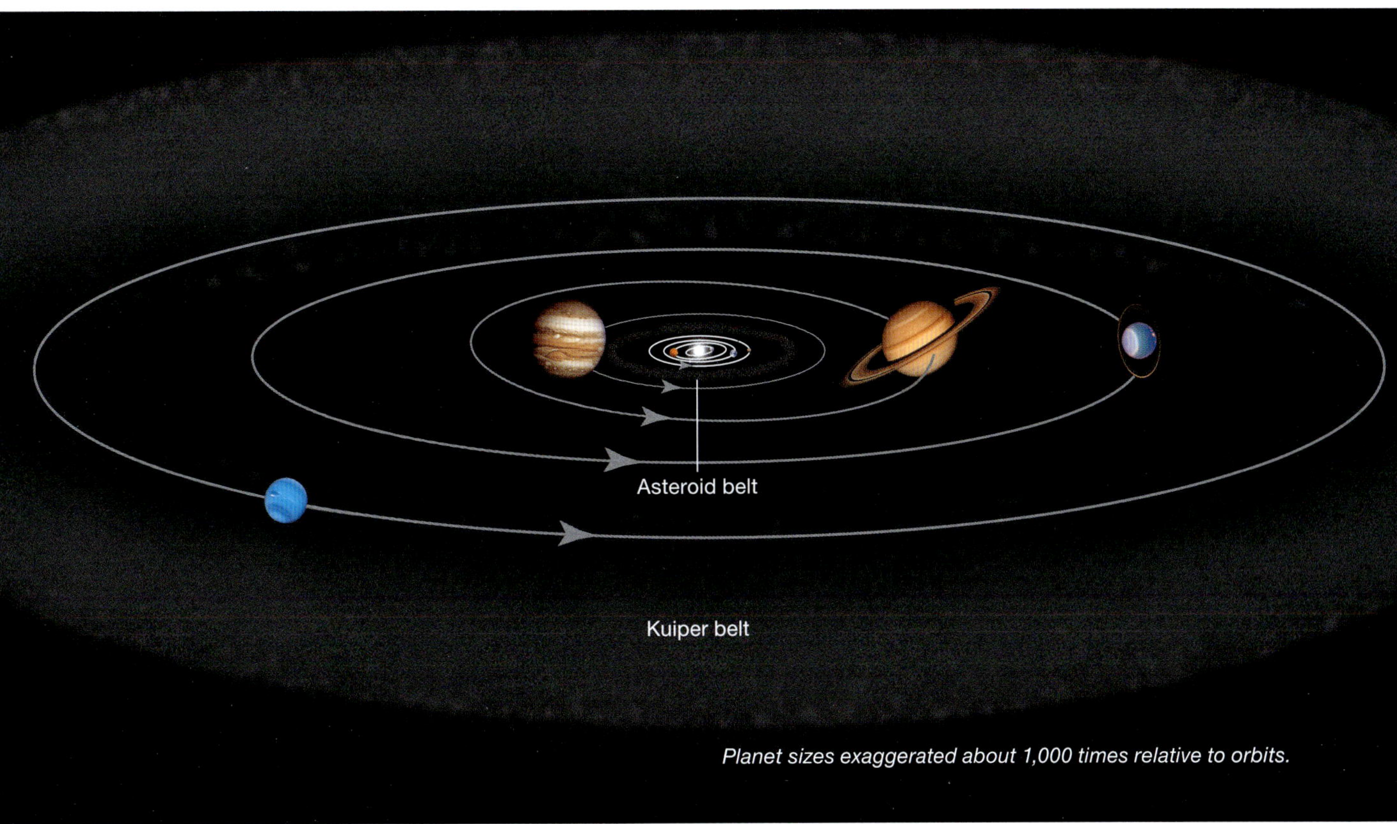

Figure 2.7. The general layout of our solar system. All planets orbit the Sun in the same direction and in approximately the same plane, as do most asteroids and most comets of the Kuiper belt. Orbits are shown roughly to scale, but the Sun's size is exaggerated by about 50 times and planet sizes are exaggerated by about 1,000 times (because they would be too small to see on the scale used for the orbits).

filament (incandescent) light bulbs give off light. The energy that keeps the Sun and other stars shining is produced deep in their cores by *nuclear fusion*, the process through which two or more small atomic nuclei fuse together to make a more massive nucleus. Our Sun produces energy by fusing hydrogen to make helium. This energy production can also be traced to the Sun's strong gravity, which compresses its core to the extreme temperatures and densities necessary for fusion.

What is a moon?

A moon is generally defined to be an object that orbits a planet. However, as with many terms in astronomy, the definition is a bit fuzzy, and we often also use the term *moon* for any smaller object that orbits a larger one, such as objects that orbit dwarf planets (for example, Pluto has five known moons) or smaller asteroids that orbit larger ones. In fact, the number of known "moons" orbiting dwarf planets and smaller objects is now larger than the number known to be orbiting planets. The term *satellite* is sometimes used as a synonym for moon, though some scientists prefer to reserve this term for human-made objects in space.

Among the eight planets, only Mercury and Venus have no moons. Earth has just one moon, which we call "the Moon" to distinguish it from others. Mars has two very small moons (named Phobos and Deimos). The outer planets each have many moons, ranging from 16 known for Neptune to 274 known for Saturn (as of 2025). Most of these moons are quite small, and it is likely that the outer planets have many more small moons yet to be discovered. However, the large moons of the planets (those large enough for gravity to make them round) have almost certainly all been discovered by now.

In case you are wondering, our Moon is the fifth largest moon in the solar system. The larger ones are Jupiter's moons Ganymede, Callisto, and Io and Saturn's moon Titan.

What are asteroids and comets?

Asteroids and comets are objects that orbit the Sun but that are too small to count as planets. As noted earlier, we distinguish them primarily by composition: Asteroids are rocky, and comets are icy. Of course, nature doesn't always separate objects as neatly as our human categories, and scientists have found many small bodies with a somewhat "in between" nature. Adding to the confusion, if an asteroid or comet is large enough to be round, we also call it a dwarf planet. For example, the dwarf planet Ceres is also known as the largest asteroid, and the ice-rich compositions of the dwarf planets Pluto and Eris mean we can also think of them as the two largest known comets.[TN17]

On a deeper level, scientists now understand that asteroids and comets are "leftovers" from the formation of the solar system. According to current scientific understanding, the planets (and major moons) formed as smaller chunks of rock and ice were gravitationally pulled together to make larger objects. Asteroids and comets represent chunks that still remain from this early time because they never got pulled into a planet. [TN18] Their leftover status means that asteroids and comets hold clues to exactly how and when the solar system formed, which explains why scientists are so interested in studying them (and have sent numerous spacecraft to explore them).

Why do all the planets orbit the Sun in the same direction?

To understand the answer to this question, we must investigate how our solar system formed. By carefully studying evidence both from our own solar system and from others— especially those that we observe to be in

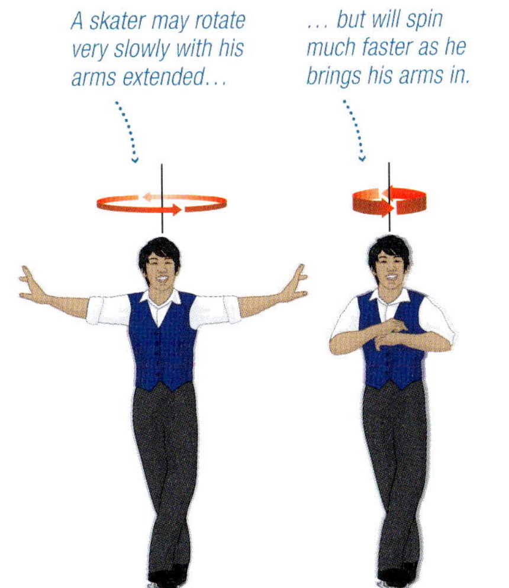

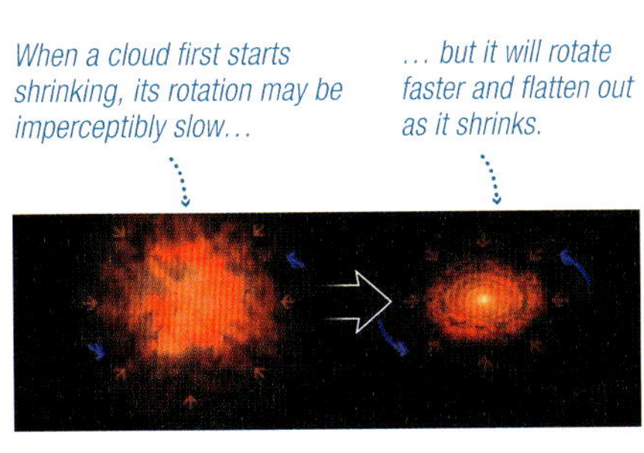

Figure 2.8. Just as an ice skater spins faster as he pulls in his arms (left), a cloud of gas spins faster as it shrinks (right). The faster rotation also causes the cloud to flatten out. Credit: Adapted from *The Cosmic Perspective*.

the process of formation today — scientists have learned that star systems are born from giant clouds of gas in space (*interstellar clouds*).

The gravity of all the gas within a cloud causes the cloud to contract in size. Like an ice skater pulling in his arms as he spins, the cloud rotates faster and faster as it shrinks (**Figure 2.8**). This rotation, in turn, causes the cloud to flatten into a disk (much as a spinning ball of dough flattens out to make a pizza crust).[**TN19**] The end result is that the gas becomes densely concentrated at the center, where a star forms (the Sun in our case). The planets form in the surrounding disk and therefore are born with orbits that share the direction and plane of the disk's rotation.

Note that if our theory is correct in telling us that the orderly orbits are a consequence of the formation process, then we should expect to see similar orbital patterns in other star systems — and we do. Astronomers have already discovered hundreds of other star systems with multiple planets, and these planets do indeed generally share the same orbital direction and orbital plane around their star.[**TN20**]

Even more impressive evidence comes from young stars that are still in the process of formation. **Figure 2.9** shows a montage of images of such systems. In each case, the star lies at the very center and is clearly surrounded by a disk-shaped cloud of material — just as our theory tells us to expect. Indeed, it is the combination of evidence like this and of statistics based on the more than 5,000 known exoplanets that has led astronomers to conclude that the vast majority of all stars must in fact be star systems with orbiting planets, moons, and their own asteroids and comets.

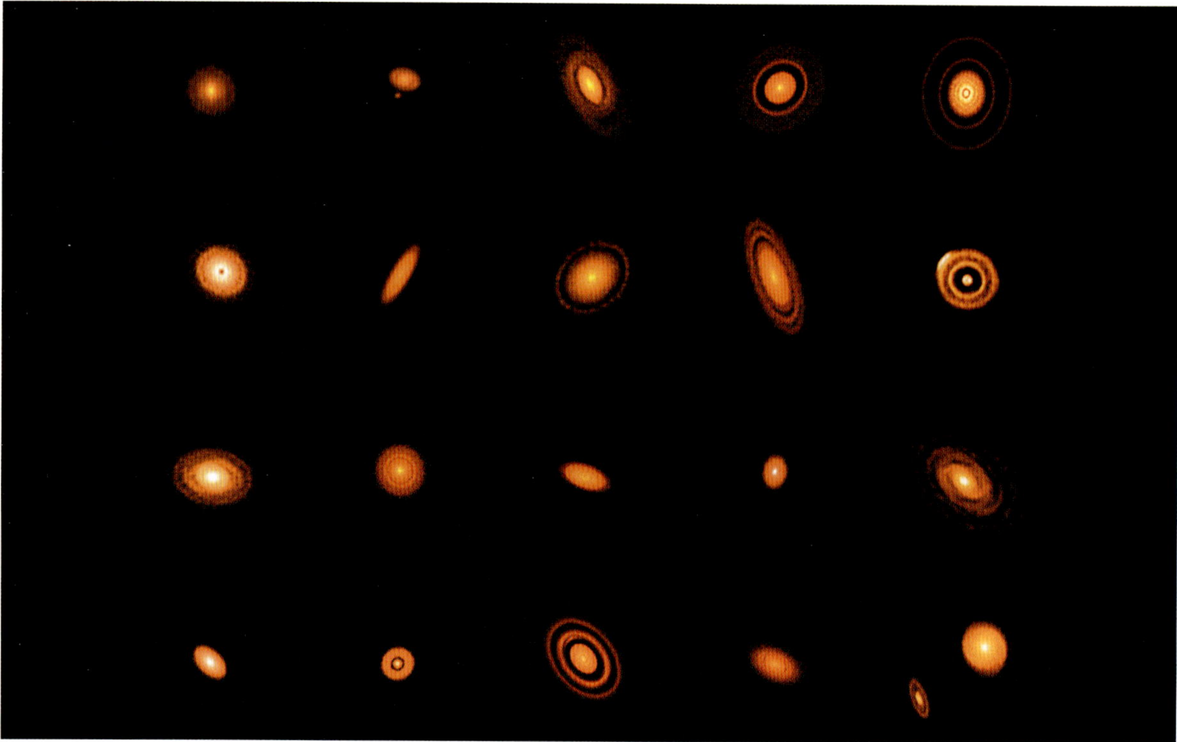

Figure 2.9. This montage shows young stars surrounded by disks of orbiting material (the star is at the center of each disk), just as our theory of solar system formation tells us to expect. In a few cases, we even see ring-shaped gaps in the disk, which are thought to be spaces where the gas has been cleared out by forming planets. Note: These images show radio wave light that our eyes could not see on their own. Credit: All images from the Atacama Large Millimeter/submillimeter Array (ALMA) in Chile.

What is a galaxy?

We now turn to the third line of our cosmic address, which states that we live in the Milky Way Galaxy. But what exactly is a galaxy?

A *galaxy* is a great island of stars in space, all held together by gravity and all orbiting a common center (the center of the galaxy). Keep in mind that when we say "stars," we presume that most of these are actually star systems with planets and other orbiting material.

Our Milky Way Galaxy is a relatively large galaxy, containing more than 100 billion star systems. We cannot see the entire Milky Way Galaxy because we live inside it. However, we can see many other galaxies that look similar. These include the relatively nearby Andromeda Galaxy, which is "only" about 2.5 million light-years away (**Figure 2.10**). Notice that the Andromeda Galaxy is shaped like a fairly flat disk with a bright central bulge. The stars in the disk are arranged in a spiral pattern, so we say that Andromeda is a "spiral galaxy." (Some galaxies have other shapes.)

▲ **Figure 2.10.** The Andromeda Galaxy (also called M31) is about 2.5 million light-years away, which makes it the nearest large galaxy to our own Milky Way Galaxy. Like the Milky Way, Andromeda is roughly 100,000 light-years in diameter. Credit: Bruce Waters/Killarney Provincial Park Observatory.jpg; used under Creative Commons license.

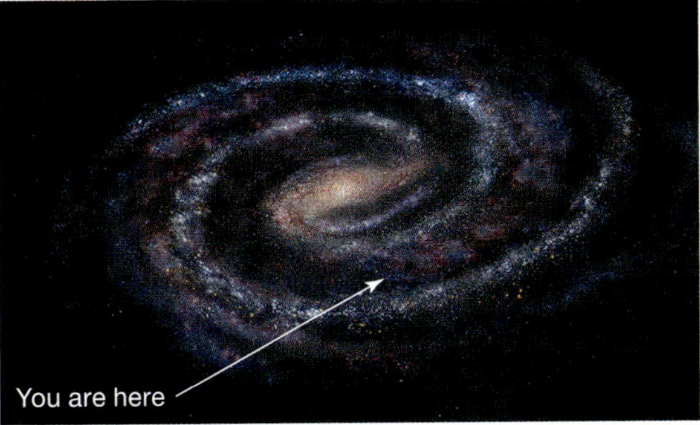

► **Figure 2.11.** This painting shows how our Milky Way Galaxy might appear if we could see it from the outside, with the approximate location of our solar system indicated. Credit: Michael Carroll.

By carefully studying the positions and motions of stars within our Milky Way, scientists have learned that our galaxy is a spiral galaxy much like Andromeda, except the central bulge is elongated into a bar shape (**Figure 2.11**). Our solar system is located about halfway out from the center of our galaxy to the edge of its disk.

In case you are wondering why we can't just send out a probe to take a picture of our Milky Way Galaxy from the outside: Our galaxy is about 100,000 light-years in diameter, meaning it takes light about 100,000 years just to cross from one side to the other. So even if we had the technology to send out a space probe at a speed close to the speed of light, it would take the probe many thousands of years just to get into a position to take a photo of our galaxy and thousands of years more for the photo to be transmitted back to us on Earth.

It's also amazing to realize that any "snapshot" of a distant galaxy is actually a picture of both space and time. For example, the fact that the Andromeda Galaxy is about 2.5 million light-years away might at first make you think that Figure 2.10 shows a snapshot of what this galaxy looked like 2.5 million years ago. However, because the Andromeda Galaxy is about 100,000 light-years in diameter, the light we are seeing from the far side of the galaxy must have left on its journey to us some 100,000 years before the light we are seeing from the near side. In other words, the photo in Figure 2.10 actually shows different parts of the Andromeda Galaxy spread over a time period of 100,000 years.

Figure 2.12. This photo shows a portion of the band of light in the night sky that we call the Milky Way. The telescopes in the foreground are part of the Cerro Tololo Inter-American Observatory (CTIO) in Chile. (Teacher Note 7 has another photo of the Milky Way.) Credit: CTIO/NOIRLab/NSF/AURA/B. Tafreshi.

Building a Cosmic Perspective Suppose that there is an advanced civilization in the Andromeda Galaxy that has an extremely powerful telescope pointed at Earth. What would they be seeing right now? Could they know that we exist?

Why is our galaxy called the Milky Way?

Our galaxy takes its name from the band of light that we also call the Milky Way, which you can see with your eyes if you go to a very dark site on a clear night (**Figure 2.12**). This band gets its name from its white, milky appearance. Our galaxy was given the same name because of its relationship to this band of light.

To understand the relationship, look back at Figure 2.11 and notice that our galaxy has a fairly flat disk, much like a pancake, and that we live inside this disk. The disk is relatively thin, and stars are widely spaced out, so we have a clear view to the distant universe when we look in directions away from the disk. However, when we look in directions *into* the disk, we are looking toward so many stars — plus interstellar clouds of gas and dust — that everything blends together into a milky white band of light. **Figure 2.13** summarizes this idea. Notice that the Milky Way band of light

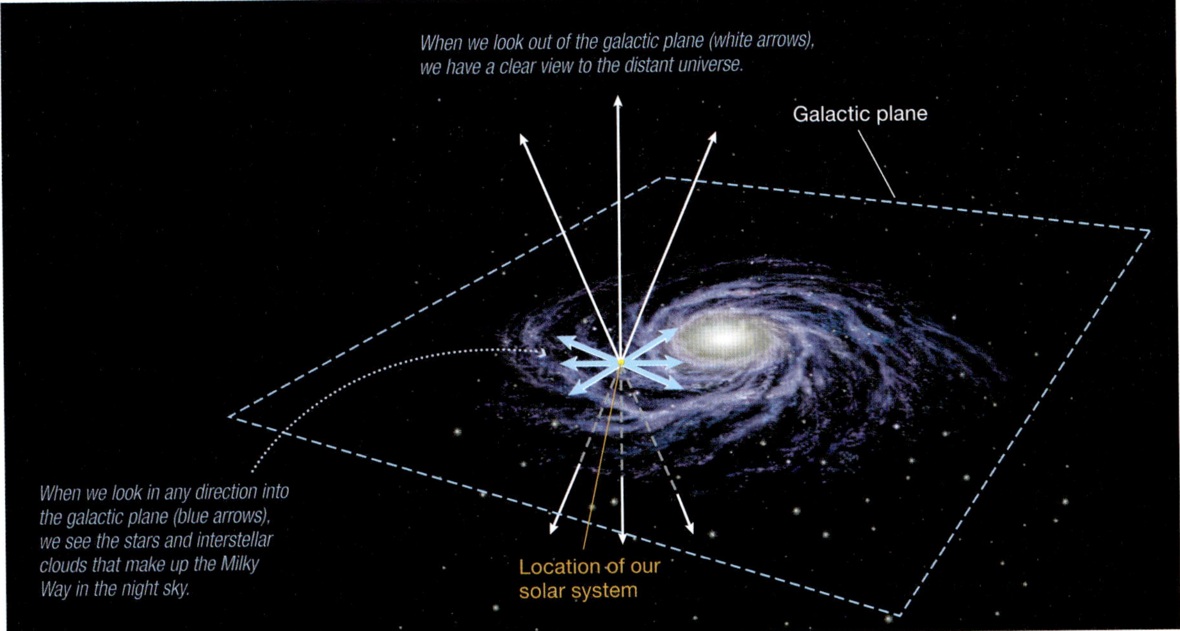

Figure 2.13. This diagram shows how our view into the disk of the Milky Way Galaxy causes us to see the whitish band of light that we call the Milky Way. Credit: *The Cosmic Perspective*.

makes a circular ring around us, which is why you can see at least some portion of this Milky Way from almost any dark location on Earth. (Note: If you look closely at the Milky Way from a dark site, you'll also see that it contains dark patches within the "milky" band; these are interstellar dust clouds that are dark because they block the light from stars behind them. Many ancient cultures, including the Incas in South America and Australian Aboriginal groups, assigned importance to the patterns of these dark patches within the Milky Way.)

What are galaxies made of?

We have said that galaxies are gigantic collections of stars (or star systems), and most galaxies also contain interstellar clouds of gas and dust. So you might at first guess that it would be reasonable to say that galaxies are made of stars and interstellar clouds. But this is only part of the story — and the full story contains one of the biggest mysteries in science.

Remember that the stars (and gas and dust) in a galaxy all orbit around the center of the galaxy, and they are held in these orbits by the gravity of the matter contained in the galaxy. Astronomers can measure the orbital speeds of the stars, and they can then use the law of gravity to calculate how much matter a galaxy must contain to hold the stars in orbit at these speeds. The results of these calculations reveal a huge surprise: For almost every galaxy that astronomers have studied, including our own Milky Way, the only way we can account for the orbital speeds is if we assume that the galaxy contains far more matter (about 10 times as much) than we would expect if we simply added up the matter in all its stars and gas clouds.

In other words, it seems that in addition to stars and clouds, galaxies must contain vast amounts of matter that we cannot see with any kind of telescope or camera. This unseen matter has been given the name *dark matter*. Dark matter is a great mystery because although the law of gravity implies that this matter must exist, scientists do not yet know what it is.

Are there structures bigger than galaxies?

Galaxies are the largest distinct collections of stars that we find in the universe, but there are even larger structures that are made up of collections of galaxies. For example, our Milky Way Galaxy is one of several dozen galaxies that make up what we call the Local Group of galaxies. (The Andromeda Galaxy is also part of the Local Group.) Some galaxy groups contain hundreds or even thousands of galaxies, in which case they are usually called *galaxy clusters*.

On even larger scales, groups and clusters of galaxies are themselves grouped together to make what we call *superclusters*. Our Local Group lies on the outskirts of what we call the Local Supercluster, or *Laniakea* (Hawai'ian for "immense heaven"). Superclusters are so enormous that we cannot easily photograph them; astronomers infer their existence by mapping the positions of galaxies in three-dimensional space. Such maps show that even superclusters are not randomly scattered through space. Instead, they form a web-like pattern like that illustrated in the background of Figure 2.1. (This pattern is sometimes called the "cosmic web.")

To summarize, while our cosmic address as given on pages 9 and 10 jumps directly from galaxy to universe, we could in principle have added two additional lines for our galaxy group and supercluster. With these additions, our cosmic address would go Earth, The Solar System, The Milky Way Galaxy, The Local Group, The Laniakea Supercluster, The Universe.

Building a Cosmic Perspective Revise your personal cosmic address (from earlier) to include our galaxy group and supercluster.

When did we learn that we live in a galaxy?

Knowing that it's been only about 400 years since we learned that Earth is a planet orbiting the Sun — and hence that we live in a *solar system* — you might wonder when we first learned that we live in a *galaxy*. The answer is surprisingly precise and recent: October 1923. That is when astronomer Edwin Hubble proved that other galaxies exist beyond the Milky Way (though the discovery was not formally announced until 1924).

Note that this was not the first time that astronomers had *seen* other galaxies. In fact, vast numbers of what we now know to be galaxies had already been photographed through telescopes long before 1923. But the telescopes of that time were not powerful enough to allow astronomers to clearly identify individual stars within other galaxies, and as a result there was great debate among astronomers about whether these objects were separate from the Milky Way or just clouds of gas within the Milky Way. In other words, we did not know whether our galaxy represented the entire universe or just one small part of it.

Hubble was working with a newly built telescope that was the largest in the world at the time (at the Mount Wilson Observatory near Los Angeles, California). This made it possible for him to identify a special type of star (called a *Cepheid variable*) within the Andromeda Galaxy. Thanks to the earlier work of an astronomer named Henrietta Leavitt, astronomers already knew that they could use this type of star to make good estimates of distance. Hubble quickly calculated that the stars he was seeing in Andromeda were so far away that they could not possibly be within our own Milky Way. With that single stroke, our perspective on the size of the universe took its largest leap ever. That is, because we now know that there are more than 100 billion galaxies in our universe, this discovery essentially taught us that the universe was at least some 100 billion times larger than we'd previously known.

What is the universe?

All the galaxies combined, including everything within and between them, make up what we call the *universe*. Another way to think of the universe is as the sum total of all matter and energy in existence.

It's worth noting that we cannot see the entire universe, even in principle. The reason is time: Based on evidence that we'll discuss in Chapter 4, scientists have discovered that the universe is about 14 billion years old. Therefore, we can only see galaxies that are near enough for their light to have reached us in 14 billion years. We presume that there are other galaxies that are more distant, but we cannot see them because their light has not yet had enough time to reach us. Scientists define the *observable universe* to be the portion of the universe that we can see.

The observable universe contains at least 100 billion large galaxies (those similar to the Milky Way in size) and many more smaller ones. To understand how we know, look back at the JWST photo that opens Chapter 1 and remember that it shows a region of the sky that you could cover with a grain of sand held at arm's length. By counting the galaxies in this photo and multiplying by the number of similar photos it would take to cover the entire sky, you can estimate the total number of galaxies in the observable universe.

Why does the universe's age determine the portion we can observe?

You can understand the answer by remembering an idea we discussed back on page 4: *The farther away we look in space, the further back we look in time.*

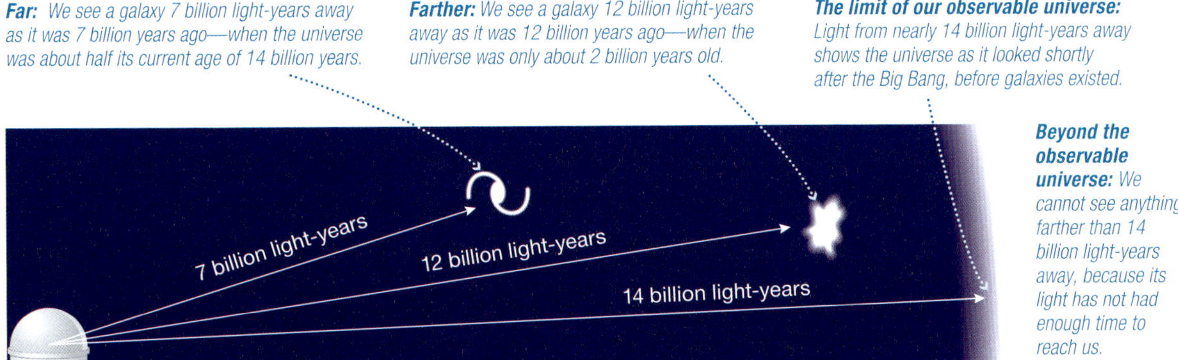

Figure 2.14. The age of the universe puts a limit on the size of the observable universe — the portion of the entire universe that we can in principle observe. Credit: *The Cosmic Perspective*.

Figure 2.14 shows how this idea limits the size of the observable universe. The first galaxy shown in the figure is 7 billion light-years away. This means we are seeing it as it was 7 billion years ago, when the universe was about half its current age of 14 billion years. The second galaxy is 12 billion light-years away, which means we are seeing it as it was 12 billion years ago — or only 2 billion years after our 14-billion-year-old universe was born. Now imagine that we tried to look at a galaxy that is 15 billion light-years away. We'd be seeing it as it was 15 billion years ago — but that was before the universe or the galaxy existed, so there is nothing to see. In other words, in a universe that is 14 billion years old, we cannot see anything that is more than 14 billion light-years away.[TN21]

What does the word *universe* mean?

The word *universe* combines the prefix *uni*, which means "one," with *verse*, which essentially means "story." In other words, the word *universe* means "one story," representing the idea that everything within it is part of the same whole.

It's worth noting that the idea of a *universe* is relatively new in human history. In ancient times, it was generally believed that Earth and "the heavens" (meaning everything we see in the sky) represented two completely different realms. Although this ancient belief was already losing favor by the time we learned (about 400 years ago) that Earth is a planet orbiting the Sun, it wasn't conclusively disproven until Isaac Newton discovered the law of gravity, which he published in the year 1687. Newton's discovery proved that the same law of gravity that keeps us bound to Earth also keeps the Moon orbiting Earth and the planets orbiting the Sun. In other words, Newton showed that Earth and the heavens were not distinct realms but instead could both be understood as part of the same "one story" — the universe.

What is a "multiverse"?

If you follow science fiction (or even some real science news), you may have heard of something called a "multiverse." No one knows if it really exists, but the idea is this: In principle, it is *possible* that there could be other universes besides the one we live in. For example, there might be universes with similar physical laws but with galaxies, stars, and people that differ from those in ours, or there might be universes with different sets of physical laws.

Keep in mind that while it can be fun (and mind-bending) to think about a multiverse, *so far there is no actual evidence that it exists*. In other words, the idea of a multiverse is purely speculative at this point. So while it's good to understand what we mean by a multiverse, there's nothing we can actually study about it at this time.[**TN22**]

Key Astronomical Definitions

This box summarizes key terms we have discussed. Note that they are not in alphabetical order but instead are organized with distance units first, followed by objects roughly in order from small to large (but grouped together in sets that make sense).

astronomical unit (AU): The average distance between Earth and the Sun, which is about 150 million kilometers (93 million miles).

light-year (ly): The distance that light can travel in 1 year, which is about 10 trillion kilometers (6 trillion miles).

asteroid: A relatively small and rocky object that orbits a star.

comet: A relatively small and ice-rich object that orbits a star.

moon (or **satellite**): An object that orbits a planet, though the term is also used for objects orbiting dwarf planets and for other smaller objects that orbit larger ones.

planet: A moderately large object (currently taken to mean at least about a third the diameter of Earth) that orbits a star and shines primarily by reflecting light from its star.

dwarf planet: An object that orbits a star and is large enough for its gravity to make it round but not large enough to count as a "full" planet. (See Teacher Note 10 for more detail on the distinction between planets and dwarf planets.)

exoplanet: A planet orbiting a star other than our Sun.

solar system: The Sun and all the material that orbits it, including the planets. The term *solar system* technically refers only to our own star system (because solar means "of the Sun"), but it is often applied to other star systems as well.

star system: A star (or sometimes more than one star) and any planets and other materials that orbit it.

star: A large, glowing ball of gas that generates heat and light through nuclear fusion in its core. Our Sun is a star.

galaxy: A great island of stars in space, containing from a few million to a trillion or more stars, all held together by gravity and orbiting a common center.

universe (or **cosmos**): The sum total of all matter and energy — that is, all galaxies and everything within and between them.

How far away do you think the Moon would be on this same scale? Before you read ahead to the answer, make a guess by trying the simple activity depicted in **Figure 3.3** and described below.

Step 1. Find a standard-size globe, which is about 30 centimeters (12 inches) in diameter, to represent Earth. Note: If you don't have a globe, a basketball is a reasonable substitute, though it is about 20% smaller in diameter than a standard globe.

Step 2. Find a ball that is roughly the correct size of the Moon on this same scale. An 8-centimeter (3-inch) Styrofoam ball works well. If you don't have a Styrofoam ball handy, you can use a baseball or even your fist to serve as an approximate representation of the Moon to scale.

Step 3. Now take your "Moon" and move it to your best guess of its correct scaled distance from your Earth globe.[TN23]

How far away did you put your Moon? If you are like most people, you put the Moon much closer to Earth than it actually belongs. To find the correct answer, you can divide the Moon's average distance

Figure 3.2. These photos show the *sizes* of Earth and the Moon to scale. Credit: Earth from NASA/Apollo 17, Moon from Gregory H. Revera, used under Creative Commons license.

of 384,000 kilometers (239,000 miles) by Earth's diameter of 12,760 kilometers (7,930 miles). The result is that the Moon's average distance is about 30 times the diameter of Earth (30 "Earth diameters"). This means that on our scale, in which Earth is a 30-centimeter (1-foot) standard globe, the Moon should be about 9 meters (30 feet) away!

If you now place your Moon at its correct scaled distance (**Figure 3.4**), you'll probably be quite surprised at how far away it is. This is likely to be only the first of many surprises you'll encounter as we continue to explore the scale of the universe.

Building a Cosmic Perspective How close was your guess to the correct answer about the Earth-Moon distance to scale? Why do you think that most people are surprised by the correct answer?

You can get a sense as to why this is so surprising by thinking about the full moon in our sky. You probably think of the full moon as looking pretty big. But if you grab a pencil and hold it at arm's length, you'll find that the eraser alone can completely cover the full moon. (Try this at the next full moon.) Interestingly, this is true at all times, no matter

Figure 3.3. You can use a standard globe and a 3-inch Styrofoam ball (or, alternatively, a baseball or your fist) to represent the sizes of Earth and the Moon approximately to scale. How far away do you think the Moon should be on this same scale? (Photo by the author.)

Figure 3.4. Here are the globe Earth and Styrofoam ball Moon separated by their correctly scaled average distance. (Photo by the author.)

whether the full moon is near the horizon or high overhead. In other words, the fact that the full moon *appears* larger when it is near the horizon is an illusion. (Do a search on "the Moon illusion" to learn why this illusion occurs.)

If we turn this same idea around, it means that our Earth also looks quite small as viewed from the Moon. This fact was first brought home by the famous "Earthrise" photo taken during the Apollo 8 mission in 1968 (**Figure 3.5**). This photo had a huge impact and is often credited with being one of the major forces that pushed people to much greater concern about the environment. It is even credited with helping to inspire the annual (April 22) celebration of Earth Day. One great way to see its profound impact is through the following words about this image from the three Apollo 8 astronauts:

> *I could put my thumb up to a window and completely hide the Earth.*
> — Apollo 8 Astronaut Jim Lovell

> *The one overwhelming emotion that we had was when we saw the Earth rising in the distance over the lunar landscape. . . . It makes us realize that we all do exist on one small globe.*
> — Apollo 8 Astronaut Frank Borman

> *We came all this way to explore the moon, and the most important thing is that we discovered the Earth.*
> — Apollo 8 Astronaut William Anders

Building a Cosmic Perspective Take a few moments to contemplate the astronaut quotes about the Earthrise image. What does it mean to you to know that, from the Moon, your thumb could cover your view of the only world on which humans have ever lived?

How far away is the International Space Station?

Hundreds of astronauts have visited the International Space Station, with new visitors typically heading there at least a couple times each year. But

Figure 3.5. This iconic photo, known as "Earthrise," was taken during the Apollo 8 mission by astronaut William Anders on Christmas Eve 1968 as he orbited the Moon with crewmates Jim Lovell and Frank Borman. Because this was the first mission to carry humans as far as the Moon, it was the first time anyone had seen our planet from such a distance. Credit: NASA/Apollo 8/William Anders.

as of 2025, only 12 people have ever walked on the Moon (all between 1969 and 1972 as part of NASA's Apollo program), though NASA is working to send people within the next few years with its Artemis program. (China also aims to send people to the Moon by 2030.) So you might wonder: Is it really that much more difficult to get to the Moon?

One way to think about the question is by considering the relative distances to scale. You have already seen that the Moon is about 30 Earth diameters away. In contrast, the International Space Station orbits Earth at an average altitude of only about 400 kilometers (250 miles) — which is barely 3% of Earth's diameter. Putting the two facts together, we find that the Moon is approximately *1,000 times as far* from Earth as the International Space Station. If you again used a standard globe for your Earth (as in the above activity), the Space Station orbit would be less than about 1 centimeter (1/3 inch) above the surface.

Keep in mind that the huge difference in distance is not the only reason why it is so much more difficult to reach the Moon. Other factors include the need to break free of Earth's gravity, which requires a higher speed; the need to have fuel to slow the spacecraft into orbit around the Moon; the need for a lander to take astronauts safely to the surface; and the need for rockets and fuel to take the astronauts back up and back home.

How thick is Earth's atmosphere on this scale?

The atmosphere does not have an abrupt end; it just gets thinner and thinner with altitude. However, about two-thirds of the air in Earth's atmosphere lies within 10 kilometers (6 miles) of the surface, which means you could represent it on a standard globe with a layer only as thick as a dollar bill.

Going higher, there is so little air by an altitude of 100 kilometers (60 miles) that this altitude is generally considered to mark the threshold of space. (This altitude is sometimes called the *Karman line*, named for physicist Theodore von Kármán.) This is only about a quarter of the altitude at which the International Space Station orbits, which means it would be less than about 3 millimeters (1/10 inch) above the surface of a standard globe.

You can see the incredible thinness of our atmosphere simply by looking at photos of our planet from space. Notice, for example, that although you can see clouds in Figure 3.5, the atmosphere seems to have no measurable thickness at all. Even a closer-up view shows our atmosphere as just a thin blue layer (**Figure 3.6**), an idea beautifully captured in another astronaut quote:[TN24]

Figure 3.6. The thin blue layer of Earth's atmosphere is illuminated by the Sun on the horizon in this photograph from the International Space Station in 2009. Credit: NASA/ISS.

> *For the first time in my life I saw the horizon as a curved line. It was accentuated by a thin seam of dark blue light — our atmosphere. Obviously this was not the ocean of air I had been told it was so many times in my life. I was terrified by its fragile appearance.*
> — Ulf Merbold, German Astronaut

How big is the Sun on this scale?

If you look up the numbers, you'll find that the Sun is more than 100 times as large as Earth in diameter (more precisely, about 109 times as large). This means that showing the Sun on our Earth/Moon scale (on which Earth is the size of a standard globe) would require a ball more than 30 meters (100 feet) in diameter. It's going to be hard to find a ball that size. Moreover, the distance from Earth to the Sun on this scale would be about 3.5 kilometers (2.2 miles). Clearly, we're going to need a new scale in order to move beyond the Earth-Moon system.

Before we do our scale jump, take a look at **Figure 3.7**, which shows Earth and the Sun to scale. Notice that Earth could easily fit inside many of the sunspots (the little black spots in the photo) that we see on the Sun. [TN25] In fact, the entire Earth-Moon system could easily fit within the Sun, because the Sun's diameter is more than three times as large as the 30 Earth diameters that separate the Moon from Earth.

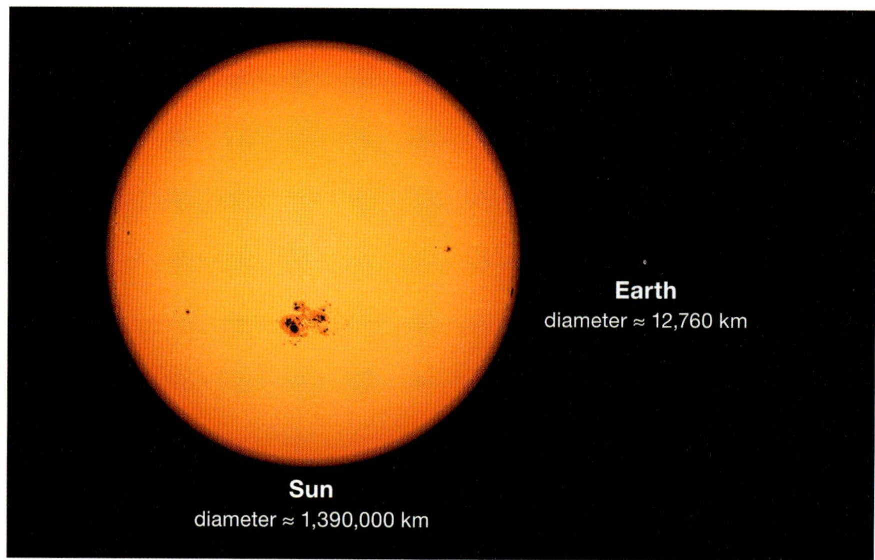

Figure 3.7. These photos show the *sizes* of the Sun and Earth to scale. Credit: Sun from NASA/SDO, Earth from NASA/Apollo 17.

Building a Cosmic Perspective A *Time* magazine cover once suggested that an "angry Sun" was becoming more active as human activity changed Earth's climate. While the Sun does indeed become more active at times (and less active at other times)[TN26], do you think that human actions could actually *cause* such changes on the Sun? Why or why not?

How big is our solar system?

We'll now jump to a scale on which it is possible to walk through our solar system. We'll use the 1 to 10 billion scale of the Voyage scale model solar system (**Figure 3.8**).[TN27] In other words, the Voyage model shows both sizes and distances in our solar system at *one ten-billionth* of their actual values. You can see in Figure 3.8 that the model Sun is about the size of a large grapefruit on this scale, which makes it easy to visualize holding the Sun in your hands. The photo also shows the pedestals for the four inner planets (Earth is the one near where the person in the blue shirt is standing); the planets themselves are encased in glass crystals within the sidewalk-facing disks visible about midway up the tall "fins."

Figure 3.8. This photo shows the pedestals housing the Sun (the gold sphere on the nearest pedestal) and the inner planets in the Voyage scale model solar system located just outside the National Air and Space Museum (Washington, DC). The model planets are encased in the sidewalk-facing disks visible above the information plaques. Note: Many other communities also have Voyage models; visit voyagesolarsystem.org for locations or to learn how you can bring a Voyage model to your own campus or community. (Photo by the author.)

Figure 3.9 shows the sizes and locations of the Sun and planets (and dwarf planet Pluto[TN28]) on the Voyage scale. Notice that the planets are quite small compared to the grapefruit-size Sun. The largest, Jupiter, is only about the size of a marble on this scale, and Earth is about the size of the ball point in a pen. The planets seem even tinier when you contrast their sizes with their distances from the Sun. For example, the ball-point-size Earth is located about 15 meters (49 feet) from the grapefruit-size Sun.[TN29]

Perhaps the most striking feature of our solar system when we view it to scale is its *emptiness*. The Voyage model has the planets arranged along a fairly straight path (for easy walkability). If we wanted to show the full orbital paths of all the planets on this same scale, we'd need an area measuring more than a kilometer (0.6 mile) on each side, equivalent to more than 300 football fields arranged in a grid (with the Sun at the center). Spread over this large area, only the grapefruit-size Sun, the planets and dwarf planets, and a few moons would be big enough to see. The rest of it would look virtually empty — that's why we call it *space*!

Building a Cosmic Perspective Take a moment to visualize this discussion. Imagine balancing a tiny ball point representing Earth on the tip of your finger. Now picture yourself standing just 15 meters (49

(a) The scaled sizes (but not distances) of the Sun and planets (and Pluto) on the Voyage scale. Note: The sizes will be nearly exact if you are reading a printed version of this book; on an electronic device, the sizes will be correct if you zoom so that the entire figure is about 7.5 centimeters (3 inches) tall. Credit: *The Cosmic Perspective*.

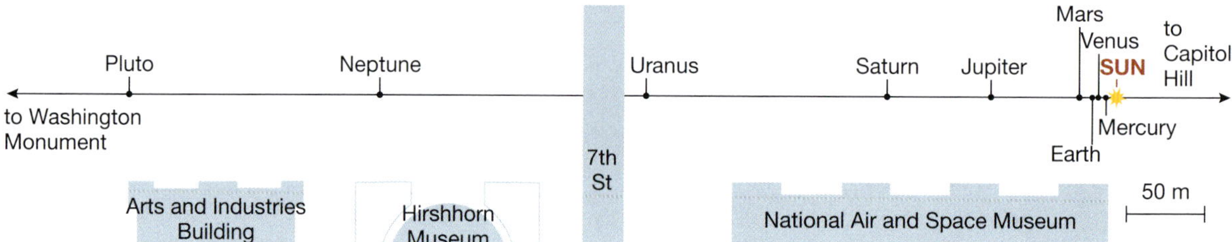

(b) Locations of major objects in the Voyage model in Washington, DC. Credit: *Voyage* project.

Figure 3.9. The Voyage scale model represents sizes and distances in the solar system at *one ten-billionth* of their actual values. Planets are lined up in the model, but in reality each planet orbits the Sun.

feet) from a grapefruit-size Sun at the center of a field the size of 300 football fields. What would your Earth look like if you could see it from the edge of the field? How does this correspond to the "pale blue dot" of Figure 3.1?

Seeing our solar system to scale also helps put space exploration into perspective. The Moon, the only other world on which humans have ever stepped, lies only about 4 centimeters (1½ inches) from Earth in the Voyage model (**Figure 3.10**). On this scale, the palm of your hand can cover the entire region of the universe in which humans have so

(a) This famous photograph from the first Moon landing (Apollo 11 in July 1969) shows astronaut Buzz Aldrin, with Neil Armstrong reflected in his visor. Armstrong was the first to step onto the Moon's surface, saying, "That's one small step for a man, one giant leap for mankind." (When asked why this photo became so iconic, Aldrin replied, "Location, location, location!") Credit: NASA.

(b) This photo shows the crystal holding the model Earth and Moon in the Voyage Scale Model Solar System located on the campus of the University of Colorado, Boulder. (Photo by the author.)

Figure 3.10. The Moon is the farthest that humans have yet traveled, but on the Voyage scale it is only about 4 centimeters (1½ inches) from Earth.

far traveled. Now contrast this Earth-Moon distance with the distances between the planets, and you can see why we've sent only robotic spacecraft throughout the rest of our solar system. Moreover, while you can walk from Earth to Pluto in only about 10 minutes on the Voyage scale, the New Horizons spacecraft, which flew past Pluto in 2015, took more than nine years to make the real journey, despite traveling at a speed nearly 100 times that of a commercial jet.

You can discover a lot of other interesting facts about the scale of our solar system if you study Figures 3.8, 3.9, and 3.10 more carefully. Even better, take a tour through a Voyage scale model (there's a list of where they are located at voyagesolarsystem.org). You can also take a virtual Voyage tour at bigkidscience.com/planetary-tour.[TN30]

Building a Cosmic Perspective We have focused on how tiny the planets appear to scale, but it's also important to keep in mind that every planet is an entire world of its own. Read through the planet pages in the virtual tour at the link above and briefly describe what you would most want to see if you could visit each of these worlds.

Why is it so much more difficult to go to Mars than the Moon?

While humans have been to the Moon (and plan to return again soon), we've sent only robotic spacecraft to Mars. There are a number of reasons why sending people to Mars will be much more difficult than going back to the Moon, but the Voyage scale makes it easy to understand one of the major reasons: A trip to Mars will almost certainly require spending at least two years away from Earth.

To see why, consider that Mars is about 8 meters (26 feet) from Earth in the Voyage model (see the locations of the pedestals for Earth and Mars in Figure 3.8). This is roughly 200 times as far as the Moon — and this is the case only because the model has Mars lined up with Earth on the same side of the Sun. Now remember that Earth and Mars both orbit the Sun, with Earth completing an orbit every one year and Mars taking a little under two years (1.88 years, to be more precise).

Let's use these facts to imagine a roundtrip journey to Mars. For the fastest possible journey, you would take off to Mars when it was approaching the same side of the Sun as Earth, timing your launch to account for the orbital motions of the two planets. With current rocket technology, this trip would take at least a few months; let's call it six months for simplicity. So you'd arrive on Mars about six months after leaving Earth. You'd look around, you'd explore, and let's say you then wanted to come home. There's a problem: During those six months that you traveled to Mars, Earth would have moved halfway around its orbit, so it would now be on the opposite side of the Sun, which means it would be much too far from Mars for you to come home.

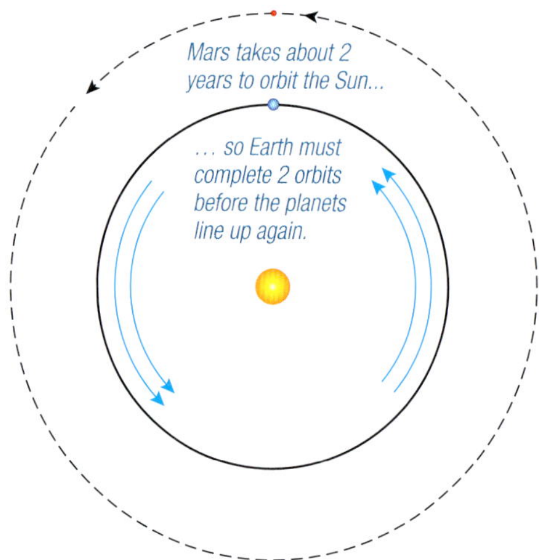

Figure 3.11. Earth and Mars line up on the same side of the Sun only about every two years (more precisely, about every 26 months).

When would Earth and Mars be positioned to make it possible for you to come home? After one year, Earth would be back at its original position, but because Mars takes close to two years to orbit the Sun, Mars would now be on the opposite side of the Sun. It's only after a second year that both planets would again be lined up on the same side of the Sun (**Figure 3.11**). (More precisely, the two planets align about every 26 months.) In other words, because you have to wait for another orbital alignment to return home, your trip to Mars is going to require that you be away for at least about two years.[TN31]

How far are the stars?

Imagine that you visit the Voyage scale model solar system in Washington, DC. Starting at the grapefruit-size model Sun, you can walk to the model Earth in a matter of seconds, and if you don't stop you can reach Pluto in only about 10 minutes. Now suppose you want to keep going beyond our solar system to walk to the nearest star (besides the Sun) on the same 1 to 10 billion scale. How far will you have to go? Make a guess before you read on.

Before we get to the answer, I'll tell you about what typically happens when I ask this question of kids at the end of a tour of a Voyage scale model solar system. Having walked the approximately 600 meters (1/3 mile) from the Sun to Pluto, most of them initially guess that they'd have to walk only a few times as far to reach the next star. When I tell them that's not far enough, they'll raise their guesses to perhaps 10 or 20 times as far. If I don't respond right away, there will be one kid who will shout out a much larger number, such as "100 miles" (160 kilometers) — and the rest of the kids will laugh, assuming that this was a ridiculously large value.

But it's not. The nearest star besides the Sun is part of the Alpha Centauri system, which is located about 4.3 light-years away.[TN32] Recall that one light-year is about 10 trillion kilometers (see page 2), so if you divide by the Voyage scale factor of 10 billion, you'll find that a light-year is about 1,000 kilometers on the 1 to 10 billion Voyage scale![TN33] Therefore, the 4.3-light-year distance to Alpha Centauri becomes about 4,300 kilometers (2,700 miles) — which is roughly the distance across the United States (**Figure 3.12**).

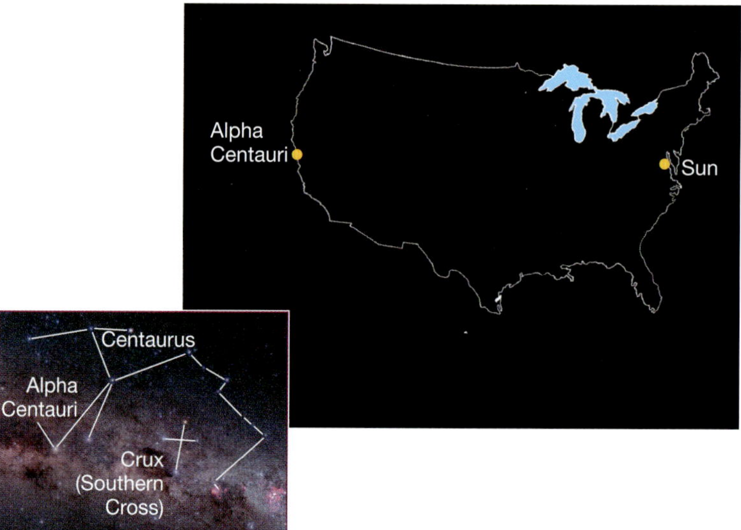

Figure 3.12. On the Voyage scale, in which the Sun is the size of a grapefruit, the distance to Alpha Centauri (the nearest other star system) is roughly the distance across the mainland United States. The inset shows Alpha Centauri's location among the constellations; it is the brightest star in the constellation Centaurus. Note: Centaurus is a southern constellation that is not visible from most of the United States. Credit: *The Cosmic Perspective*.

In other words, on the same scale with which you can walk the length of our planetary system in about 10 minutes, *you'd have to cross the United States* to reach the next star. This means that viewing Alpha Centauri in the night sky is somewhat like being in Washington, DC and seeing a very bright grapefruit in San Francisco (neglecting the problems introduced by the curvature of Earth). It may seem remarkable that we can see the star at all, but the blackness of the night sky allows the naked eye to see it as a faint dot of light.

Building a Cosmic Perspective Many science fiction shows make interstellar travel look easy. Given what you've just learned about the distances to the stars as seen to scale, comment on the level of realism (or lack thereof) in various science fiction shows that portray interstellar travel.

Now think about the challenge of detecting *planets* orbiting nearby stars, which is equivalent to being in Washington, DC and trying to find ball points or marbles orbiting bright grapefruits in California or beyond. When I wish to be amazed about modern technology, I think about the fact that we can now do this, as astronomers have already discovered thousands of planets orbiting other stars (including at least two in the Alpha Centauri system).

How long would it take us to travel to other stars?

Another way to think about the tremendous distances to stars is to consider spacecraft travel. The New Horizons spacecraft was sent into space at a speed of about 50,000 kilometers per hour (30,000 miles per hour), which is about as fast as we can achieve with current technology. Even so, New Horizons took more than nine years to reach Pluto, which it quickly flew past. It is now continuing its outward journey, but despite its high speed, New Horizons would need almost *100,000 years* to travel the distance to the nearest stars. Science fiction shows like *Star Trek* and *Star Wars* may make interstellar travel look easy, but it remains far beyond our present technology.

What does this mean for UFOs?

The tremendous distances to the stars also offer an important reality check on claims that we are being visited by aliens in UFOs. As we'll discuss later, it's not only possible but perhaps even likely that alien civilizations capable of interstellar travel exist. If they do, it's also possible that some of them might send visitors to Earth. But let's think about the technology these aliens would possess.

By now we have spacecraft photos of all the planets and major moons in our solar system, and it is quite clear that none of these worlds are home to a technologically advanced civilization. Therefore, if aliens are visiting Earth, they don't come from around here; they must come from the stars. Moreover, if you believe UFO reports, these aliens are capable of traveling among the stars easily and often.

We can put this in terms of our model solar system. Remember that, on our 1 to 10 billion scale, both Earth and the Moon fit in the palm of your hand — and the Moon is the farthest that a human has ever traveled — while the aliens must be coming from a star system that is at least thousands of kilometers away.

So there you have it: If UFO reports are to be believed, alien visitors can in essence do the equivalent of flitting back and forth across the United States while we've never left the palm of a hand. I have no idea what such technology might look like. What I *do* know is that this technology is far beyond what we have and very likely beyond what we can even yet conceive of. Indeed, I find it useful to think about this famous quote from the science fiction writer Arthur C. Clarke: "Any sufficiently advanced technology is indistinguishable from magic."

My conclusion is that if aliens really are visiting us, then to us their technology would seem almost magical. To me, this makes claims of UFO "evidence" for aliens seem far-fetched, because if they wanted us to know they were here, they'd surely be able to tell us, and if they didn't want us to know, they'd probably have ways to stay well hidden. To sum up, it's certainly possible that advanced alien civilizations exist and might even be visiting Earth; it's just that we're unlikely to find out about them unless they want us to. (Note: I acknowledge that many people have seen "unidentified" objects or phenomena in the sky; I just doubt that these things have alien origin.)

How big is the Milky Way Galaxy?

Once we get beyond the nearest stars, the 1 to 10 billion scale of the Voyage scale model solar system is no longer useful. After all, we've already found that the distance from the Sun to the nearest stars on this scale is the distance across the United States, so there's not enough room on Earth to go much farther.

To visualize the entire Milky Way Galaxy, we'll "scale jump" by another factor of 1 billion (to a scale of 1 to 10^{19}). On this new scale, each light-year becomes 1 millimeter, and the 100,000-light-year diameter of the Milky Way Galaxy becomes 100 meters, or about the length of a football field (**Figure 3.13**). With the galaxy centered over midfield, our solar system would fit within a microscopic dot around the 20-yard line. The 4.3-light-year separation between our solar system and Alpha Centauri becomes just 4.3 millimeters (1/6 inch) on this scale — smaller than the width of your little finger. If you stood at the position of our solar system in this model, *millions* of star systems — including nearly all of the stars that our eyes can see in the night sky[**TN34**] — would lie within reach of your arms.

Of course, visualizing the galaxy on a football field can tend to make you forget the enormous scale involved. One way to keep an appropriate perspective is to recognize that because the full football field represents the Milky Way's diameter of about 100,000 light-years, our location near the 20-yard line means a real distance of about 80,000 light-years to the far side of the galaxy. In other words, if we sent a radio

Figure 3.13. This painting shows the Milky Way Galaxy on a scale where its diameter is the length of a football field. The "you are here" arrow represents our solar system's approximate location on this scale. Credit: Painting by Michael Carroll.

message (which travels at the speed of light[TN35]) today to someone living on the far side of our own galaxy, they wouldn't receive the message until some 80,000 years from now — and it would be 160,000 years until we'd receive any response they might send.

Another way to put our galaxy's immensity into perspective (and my personal favorite) is to think about its number of stars. Imagine that you are having difficulty falling asleep tonight (perhaps because you are contemplating the scale of the universe) and instead of counting sheep, you decide to count stars. If we conservatively assume that our galaxy contains 100 billion stars, how long would it take you to count them?[TN36]

Building a Cosmic Perspective Before you continue, be sure to make a guess about how long it would take you to count 100 billion stars.

If you could count them at a rate of one per second, then it would obviously take you 100 billion seconds. But how long is that? A quick story before we get to the answer: When I give this question to children, they often object that they can count faster than one per second and will demonstrate by counting rapidly from 1 to 10. I then point out that it will be much more difficult to keep up the pace when they get to, say, "sixty-two billion, four hundred seventy-nine million, three hundred eighty-one thousand, five hundred forty-seven" (and will they be able to keep track of what comes next?). So a counting pace of one per second would actually be pretty ambitious.

Getting back to the question, you can find the answer by dividing 100 billion seconds by 60 seconds per minute, 60 minutes per hour, 24 hours per day, and 365 days per year. If you do this calculation, you'll find that 100 billion seconds is nearly 3,200 years. In other words, you would need thousands of years just to *count* the stars in the Milky Way Galaxy, let alone to give them names, study them, or search their planets for signs of life. And this assumes you never take a break—no sleeping, no eating, and absolutely no dying!

You might now try to combine the above fact with what you learned earlier about the separation between our solar system and Alpha Centauri (which is typical for the separations of stars in much of our galaxy). That is, try to put the following two facts together: (1) It would take you thousands of years to count the stars in our galaxy, and (2) each pair of stars is typically separated by a distance that, to scale, is like grapefruits on opposite sides of the United States. I'm not sure it's possible for the human brain to truly comprehend both facts at once, but thinking about them at least gives a little perspective on the immensity of our galaxy.

How many *planets* are there in the Milky Way Galaxy?

As we discussed earlier (see page 5), statistical evidence suggests that most stars have planets. Therefore, since we are only being approximate in saying that the galaxy has 100 billion stars, we can also say that our galaxy has about 100 billion star systems — meaning systems with a star (or sometimes more than one star) and planets. If we assume that our solar system is typical with its eight planets, then the total number of planets would be somewhere in the 800 billion range (and could be much higher, since 100 billion is a very conservative approximation for the actual number of stars in our galaxy).

What about the number of planets that might be "Earth-ish" enough to be potential homes for life? Per our earlier discussion, it seems reasonable to think that, on average, there might be about one such planet per star system. Therefore, when we say that there are about 100 billion stars in the Milky Way Galaxy, we could just as easily say that there are probably about 100 billion "Earth-ish" planets — which means that it would also take you thousands of years to count these possible life-bearing worlds.

How big is the universe?

As incredible as the scale of our galaxy may seem, the Milky Way is only one of more than 100 billion galaxies in the observable universe. Just as it would take thousands of years to count the stars (or planets) in the Milky Way, it would take thousands of years to count all the galaxies.

We can put these ideas together to estimate the total number of stars (and planets) in all these galaxies. If we assume 100 billion stars per galaxy, the total number of stars in the observable universe is roughly 100 billion × 100 billion, which is 10,000,000,000,000,000,000,000 (10^{22}).

How big is this number? Imagine visiting a beach and running your hands through the fine-grained dry sand. Imagine counting each tiny grain of sand as it slips through your fingers. Now imagine counting every grain of dry sand on the beach, then continuing on to count every grain of dry sand on every beach on Earth. If you could actually complete this task, you would find that the number of stars in the observable universe is similar to or larger than the number of grains of sand *on all Earth's beaches combined* (**Figure 3.14**).

Building a Cosmic Perspective This fact will become even more meaningful if you actually run your hands through some sand and look at the tiny, individual grains. Try it; you can use sand at a beach, or a playground, or a sand bag.

Figure 3.14. The number of stars in the observable universe is similar to or larger than the number of grains of sand *on all Earth's beaches combined.* (Photo by the author.)

Your mind may be spinning as you think about the implications of the above fact. We'll talk about some of the most important implications momentarily, but let's first just consider what it means for the possibility of other advanced civilizations.

From a purely scientific viewpoint, we do not yet have any evidence for the existence of other civilizations, which means it's possible that we are alone in the universe. But if this is the case, it in essence means that out of all the grains of sand on all the beaches on Earth combined, there was somehow just one that was different enough from all the others that it, and it alone, gave birth to an intelligent species. Again, it's possible, but when put in those terms, it seems far more likely that we share the universe with many others.

What about the scale going smaller, toward atoms?

This book focuses on large scales, going from Earth to the universe as a whole. But you might wonder what happens as we go to smaller scales, such as to the scale of the atoms that make up our bodies and our planet. These scales are equally mind boggling. Just to give a few examples:

- Atoms are so small that it would take close to 10 million of them to span the diameter of the period at the end of this sentence.
- The number of atoms in an average human body is roughly 10^{28} — which is a factor of a million larger than the number of stars in our observable universe. In other words, the number of atoms in our bodies is about a million times as large as the number of grains of sand on all the world's beaches combined.
- The nucleus at the center of an atom, which contains virtually all of an atom's mass, is about 100,000 times smaller than the atom itself. This means that on a scale that makes an atomic nucleus the size of your fist, the full atom would be roughly a kilometer (0.6 mile) in diameter.

If you'd like to explore the scales of the small and large at the same time, a great place to start is with a classic short film called *Powers of Ten* (1977), made by Charles and Ray Eames. There are also many similar videos that have been made more recently, but it's hard to beat the original.

What does all this mean about *our* place in the universe?

The first reaction that most people have after this discussion of the scale of the universe is something like "Wow, we are so small." This is certainly true, at least when we consider the physical size of Earth and humanity in our incredibly vast cosmos. But small does not mean insignificant, and I believe that what we have discussed makes us *very* significant in at least two important ways:

- The fact that we have been able to learn all this about the universe is proof that our minds have great power. Indeed, I'd like everyone, especially kids, to internalize this lesson by always thinking: "I may be physically small, but when I put my mind to it, I am capable of great things."[**TN37**]
- Through us, the universe knows that it exists. That is, while the physical universe cannot think for itself, intelligent beings like us (and perhaps others) are part of the universe, and this means that our

knowledge also belongs to the universe as a whole. While the idea is somewhat philosophical, I think it is fair to say that, in a very profound way, *we represent self-awareness for the universe.*

Our exploration of the scale of the universe also carries another important lesson: *We are not the center of the universe.* Earth is not in the center of our solar system, our solar system is not in the center of our galaxy, and our galaxy is just one among the scattered billions of our observable universe. These facts may sound obvious today, but for most of human history it was commonly assumed that Earth was at the center of the universe. Many people took this to further imply a central position for humanity, for their local tribe, or even for themselves personally. I would argue that most of the problems we face in the world today are caused by people who have not yet taken our new understanding to heart. That is, they suffer from "center of the universe syndrome," which I define as follows:

> *center of the universe syndrome (COTUS)*: when people behave as though they believe the universe revolves around them.

Getting over center of the universe syndrome is not easy. We're all born with it, and it's human nature to want to keep thinking of oneself as somehow special, better, or more important than others. But I believe that if everyone learned to appreciate the true scale of the universe, we'd all have an easier time accepting that no person is any more central than any other and that our planet is a small and fragile oasis in the vast emptiness of space. Once we accept those things, it's much easier to recognize the importance of taking good care of our planet and of treating *all* of our fellow humans with empathy, kindness, and respect. In other words, learning about the scale of the universe is much more than just an exercise in science. It is a pathway to building a better world for generations to come.

These same ideas have been eloquently expressed by many others before, including the person who made what was probably the first reasonably accurate estimate of distances to the stars — the Dutch astronomer Christiaan Huygens (1629–1695). Here's what he said as he began to grasp the small size of our Earth in the cosmos:

How vast those Orbs must be,
and how inconsiderable this Earth,
the Theatre upon which all our mighty Designs,
all our Navigations,
and all our Wars are transacted,
is when compared to them.
A very fit consideration,
and matter of Reflection,
for those Kings and Princes
who sacrifice the Lives of so many People,
only to flatter their Ambition in being Masters
of some pitiful corner of this small Spot.
— Christiaan Huygens, c. 1690

I'll close this chapter by urging you to take another look back at our "pale blue dot" in Figure 3.1. I hope that our discussions of the scale of space have given you some new perspectives on yourself, on your species, and on your planet.

Building a Cosmic Perspective Astronomer Carl Sagan (1934–1996), who suggested having the Voyager 1 spacecraft take the "pale blue dot" image, also expressed these ideas with great eloquence. Listen to his 4-minute Pale Blue Dot speech, to which I've posted a link at bigkidscience.com/Sagan. What do you think as you listen to Sagan's words?

---

star's life and processes that occur during and after stellar death). In other words, we and our planet are made of "star stuff" produced by stars that lived and died long ago and that became incorporated into our solar system through cosmic recycling.

This brief overview obviously leaves out many details, but it should be enough to give you context for thinking about the scale of time.

Building a Cosmic Perspective The fact that we are made of "star stuff" also implies that most of the atoms from which we and Earth are made were once inside a star that lived and died long ago. This means that we are connected to the stars in a far more profound way than most of our ancestors ever imagined. What kinds of philosophical implications do you find in the fact that we are made of "star stuff"?

What exactly do we mean by the *Big Bang*?

The term *Big Bang* might seem to suggest some type of explosion, but that is not how it is understood by scientists. Rather, the Big Bang is simply the name we give to the event marking the instant at which the expansion of the universe began.

How do we know the universe is expanding?

As we've discussed, the basic idea for the Big Bang arises from the fact that we live in an expanding universe. But how do we know this? The answer comes from two key observations first made in the 1920s by Edwin Hubble and confirmed over and over ever since:[TN42]

1. Except for very nearby galaxies (those in the Local Group), all galaxies in the universe are moving *away* from us.
2. The more distant the galaxy, the faster it is racing away from us.

While these two observations might at first make it sound as if we suffer from a cosmic case of chicken pox, there is a much more natural explanation: The entire universe is expanding. **Figure 4.3** uses a simple analogy to show how the observations lead to this conclusion.

Imagine that you are a meticulous baker and you make a raisin cake in which the distance between adjacent raisins is always 1 centimeter (as shown at the top of Figure 4.3). You then place the cake in an oven, where

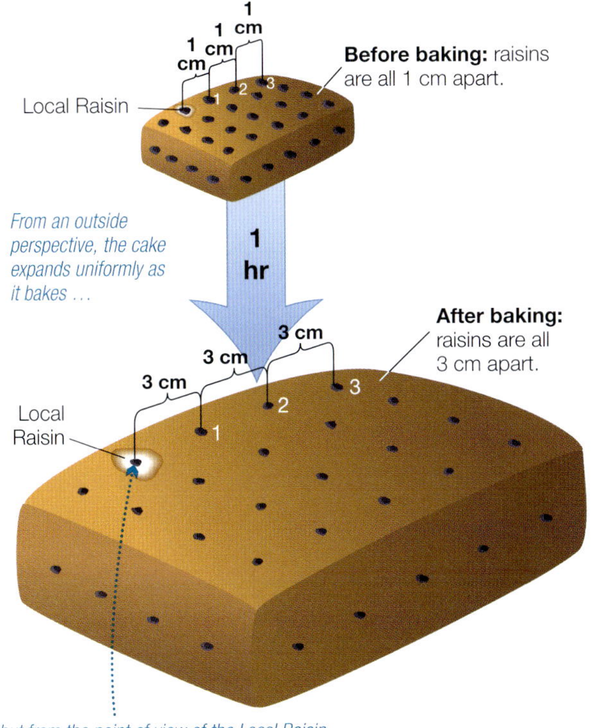

Distances and Speeds as Seen from the Local Raisin

Raisin Number	Distance Before Baking	Distance After Baking (1 hour later)	Speed During Baking
1	1 cm	3 cm	2 cm/hr
2	2 cm	6 cm	4 cm/hr
3	3 cm	9 cm	6 cm/hr
⋮	⋮	⋮	⋮

Figure 4.3. An expanding raisin cake offers an analogy to the expanding universe. Credit: *The Cosmic Perspective*.

it expands as it bakes. After 1 hour, the cake has expanded so that the distance between adjacent raisins has increased to 3 centimeters (bottom of Figure 4.3). As the figure shows, the expansion of the cake is fairly obvious from the outside. But what would you see if you lived *inside* the cake — say, within the Local Raisin (analogous to our Local Group of galaxies) — as we live in the universe?

The table accompanying Figure 4.3 summarizes what you would observe from the Local Raisin as the cake expanded. Notice, for example, that Raisin 1 starts out 1 centimeter from the Local Raisin (before baking) and ends up 3 centimeters away (after baking), which means it moves a total of 2 centimeters away from the Local Raisin during the hour of baking. Hence, its speed as seen from the Local Raisin is 2 centimeters per hour. Raisin 2 moves from 2 centimeters to 6 centimeters, which means it moves a total of 4 centimeters away from the Local Raisin during the hour. Hence, its speed as viewed from the Local Raisin is 4 centimeters per hour. Raisin 3 moves from 3 centimeters to 9 centimeters (a distance of 6 centimeters), so its speed as viewed from the Local Raisin is 6 centimeters per hour. The same pattern would continue for more distant raisins.

In other words, if you lived in the Local Raisin, you'd see all other raisins moving away from you as the cake expanded, with more distant raisins moving away faster. Moreover, you can confirm for yourself that you'd get the same results (all other raisins moving away, with more distant ones moving faster) no matter which raisin you chose as the Local Raisin. *Notice that these results are exactly what we observe for galaxies in the universe.* The obvious conclusion is that, much as we would find from inside the cake, we live within a universe that is expanding with time.

What is the universe expanding into?

The idea that we live in an expanding universe inevitably raises questions such as "What is the universe expanding into?" or "Where is the center of the universe?" Modern science tells us that the answer to the first question is that the universe is not expanding into any pre-existing space; rather, it is *space itself* that is growing with time. The answer to the second question is that the universe has *no center* (and no edges). In other words, you'd see exactly the same effects of expansion (virtually all other galaxies moving away from yours, with more distant galaxies moving away faster) no matter where you lived in the universe.

Of course, these answers probably leave you somewhat unsatisfied, since you may still wonder how the universe can be expanding without expanding "into" anything and how it's possible to have a universe with no center. Understanding the full answers to these questions would require that you first learn what Einstein's theory of relativity tells us about the nature of space, time, and gravity.[**TN43**] However, you can gain some understanding with another analogy. This time, instead of imagining an expanding raisin cake, visualize an expanding balloon, with galaxies represented by plastic dots on the surface of the balloon (**Figure 4.4**).

If you pick any particular dot to represent your Local Dot and then observe how others appear to move as the balloon expands, you'll find the same basic results that we found for the raisin cake (and that we observe for the universe): all other dots moving away from you, with more distant dots moving faster. But in this case, it's clear that the surface of the balloon is not expanding into any pre-existing surface; rather, it is the surface itself that expands with time. We mean much the same thing when we say that space itself is expanding in the universe, except it is happening in one more dimension (because a surface is two-dimensional while space is three-dimensional). In other words, in this analogy, it is the balloon's *surface* that represents space in our universe.

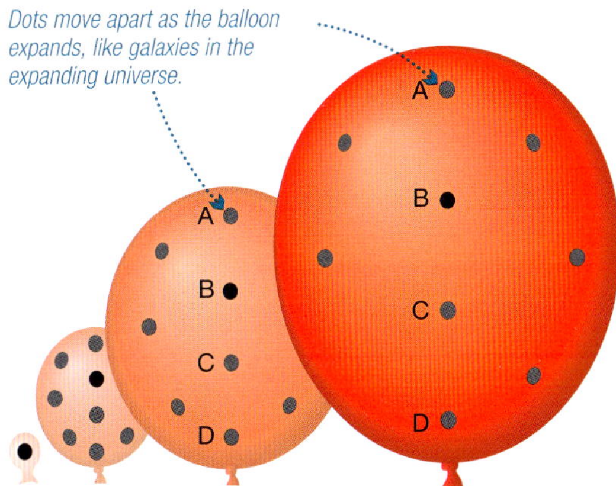

Figure 4.4. If you lived in one dot on the expanding surface of a balloon, you'd get the same basic results that we found for the raisin cake: other dots moving away, with more distant ones moving faster. But this analogy is a better one for the universe, because like space, it is the surface itself that expands with time, and like the universe, the surface of the balloon has no center (or edges). Credit: Adapted from *The Cosmic Perspective*.

The balloon analogy also shows how it is possible for the universe to have no center and no edges. That is, just as the surface of Earth has no center and no edges (for example, no particular city is any more "central" than any other and there is no edge where you would fall off the Earth), the surface of the balloon has no center and no edges. Because the balloon's surface represents the universe in this analogy, we would similarly find that the universe has no center and no edges.

How is the expansion rate changing with time?

If you throw a ball upward, gravity makes it slow down as it rises. We might similarly expect the mutual gravitational attraction of all the galaxies in the universe to slow the expansion rate with time. However, measurements made since the 1990s (largely with the Hubble Space Telescope) show the opposite: The expansion rate has been *increasing* with time, at least for the past few billion years.

What could be causing this surprising acceleration of the expansion? No one knows, but it is generally presumed that there must be some type of energy — usually called *dark energy* — that can push galaxies apart. While you might sometimes hear proposed explanations for dark energy (for example, it is sometimes attributed to a "cosmological constant"), the truth is that dark energy is simply a name given to something that no one yet understands. If you recall our earlier discussion of *dark matter* (see page 24), you'll realize that, together, dark matter and dark energy represent two of the biggest mysteries in science.

Building a Cosmic Perspective Astronomers can observe the effects of dark matter and dark energy and thereby include them in an inventory of all the known matter and energy in the universe. The results indicate that, together, dark matter and dark energy represent about 96% of the total contents of the universe — which means we don't actually know what 96% of our universe is made of. What does this fact tell you about how much we still have to learn?

What does the accelerating expansion mean for the fate of the universe?

Before the acceleration of the expansion was discovered, astronomers generally presumed that gravity must be slowing the expansion with time, in which case the universe seemed to have two possible fates: (1) If the total strength of gravity was great enough, the expansion would eventually halt and reverse; (2) otherwise, the expansion would continue to slow but never completely cease. The acceleration obviously throws a wrench into these ideas since it suggests a third possibility: that the expansion will continue forever at an ever-increasing rate.

Of course, before you assume that any of those three possibilities actually *is* the fate of the universe, it's important to remember that forever is a very long time! Just as the acceleration of the expansion came as a great surprise to astronomers when it was first discovered in the late 1990s, we may discover other great surprises between now and forever that may again change our ideas about possible fates of the universe.

What evidence tells us that the Big Bang really occurred?

As we've discussed, the fact that the universe is expanding implies that galaxies must have been closer together in the past, and running this idea backward in time suggests that the expansion must have had a beginning in the Big Bang. But this type of logical implication is not enough to constitute evidence in science. The real reason why astronomers are confident in the theory that the universe began with the Big Bang comes from three major lines of evidence, each of which verifies an important prediction of the theory:

1. *Observation of the cosmic microwave background*. Just as compressing gas inside a gasoline car engine (the piston compresses gas in a cylinder) makes the gas hotter and denser, the universe must have been hotter

and denser when it was smaller (in the past). At the time of the Big Bang, the universe must have had all its matter compressed to extremely high temperature and density, producing intensely bright radiation (light). Calculations show that as the universe expanded and cooled with time, it should have left behind a faint "glow" of radiation. This radiation, known as the *cosmic microwave background*, has been detected and studied (**Figure 4.5**), providing strong evidence that the Big Bang really occurred.

2. *Elemental abundances*. Scientists can use calculations of temperatures and densities after the Big Bang to predict exactly when and how the chemical elements should have been born in the very early universe. (Amazingly, this element formation would have occurred within the first 5 minutes after the Big Bang.) These calculations predict that if the Big Bang really occurred, then the chemical composition of the early universe should have been three-fourths hydrogen and one-fourth helium (by mass), with no other elements aside from a trace of lithium. (Recall that all other elements were made later by stars, which is why we are "star stuff.") Observations of stars and galaxies that formed early in the history of the universe confirm that the universe did indeed start with this composition. This excellent agreement between prediction and observation strongly supports the Big Bang theory.

Figure 4.5. This image from the Planck spacecraft shows an all-sky view (imagine it wrapping into a sphere around your head) of the cosmic microwave background, which is the remnant radiation of the Big Bang. (This microwave radiation is not visible to our eyes.) Credit: ESA and the Planck Collaboration.

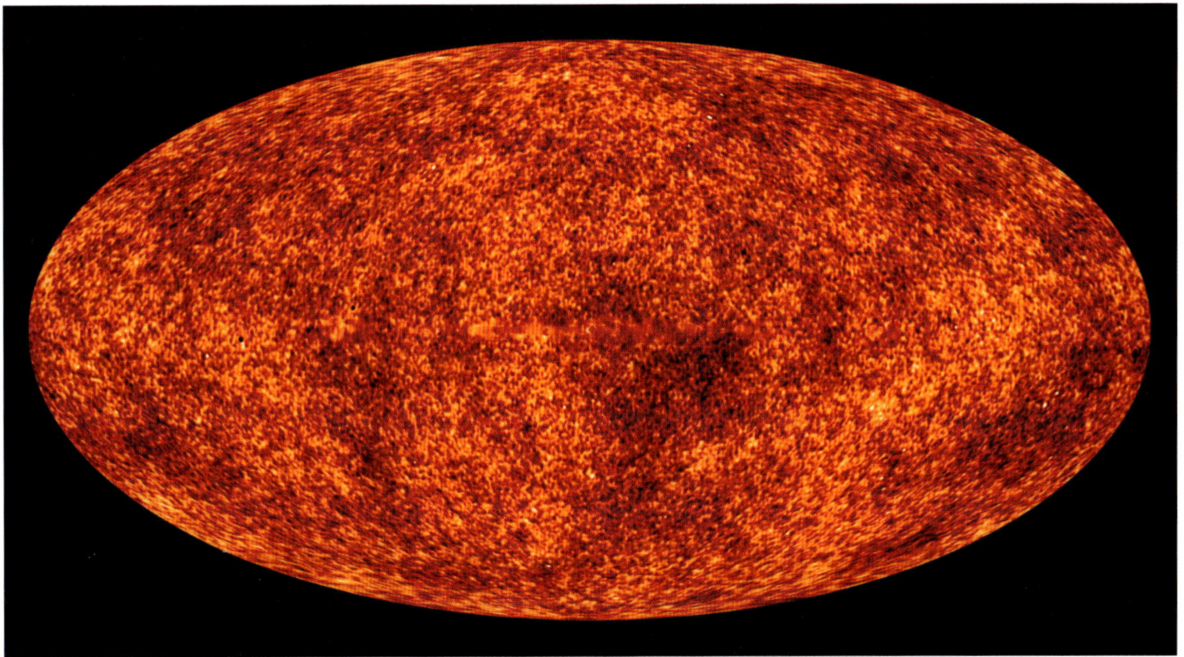

3. *Direct observation of galaxy evolution.* The idea that the universe began in a Big Bang also implies that the universe should evolve with time. For example, galaxies should have begun to form when the universe was young and then changed with time as the universe aged. Because we see more distant galaxies as they were further in the past, we can actually observe this galactic evolution through telescopic studies of galaxies at different distances. In other words, we can directly observe how the universe has changed as it has aged, providing further confirmation that it had a beginning in the Big Bang.

Together, these three lines of evidence give scientists great confidence not only that the Big Bang occurred but also that we have a strong understanding of exactly what unfolded from a fraction of a second after the Big Bang to the present day.

Where do we fit into the scale of time?

We are now ready to put the 14-billion-year history of the universe into perspective. We'll use a scaling method called a "cosmic calendar" (popularized by Carl Sagan), in which we imagine compressing the entire history of the universe into a single year. In other words, the Big Bang takes place at the first instant of January 1 and the present is the stroke of midnight on December 31. For a universe that is about 14 billion years old, each month on the cosmic calendar represents a little more than 1 billion years, each day represents about 40 million years, and every second represents more than 400 years.

Figure 4.6 shows our cosmic calendar. Most galaxies, including our own Milky Way, probably took shape by February. Many generations of stars lived and died in the subsequent cosmic months, enriching our galaxy with the "star stuff" from which we and our planet are made.

Our solar system and our planet did not form until early September on this scale, or about 4½ billion years ago in real time.[TN44] We do not yet know exactly when life on Earth began, but it was clearly flourishing by late September on this time scale. However, for most of Earth's history, living organisms remained microscopic (though sometimes living in large groups, such as those that created the fossils called *stromatolites*). On the scale of the cosmic calendar, recogniz-

THE COSMIC CALENDAR

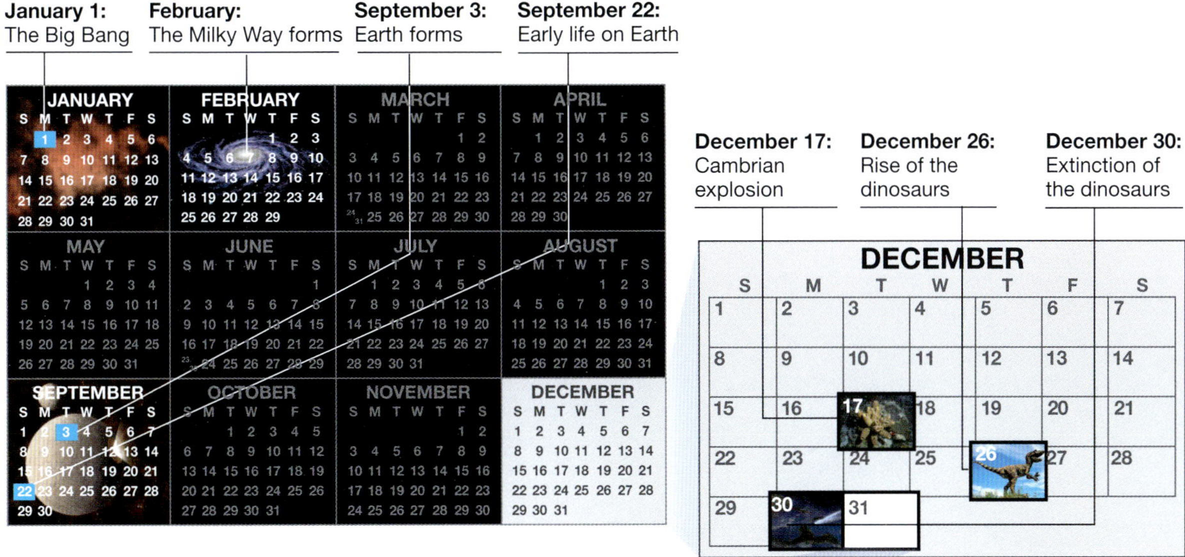

December 31:

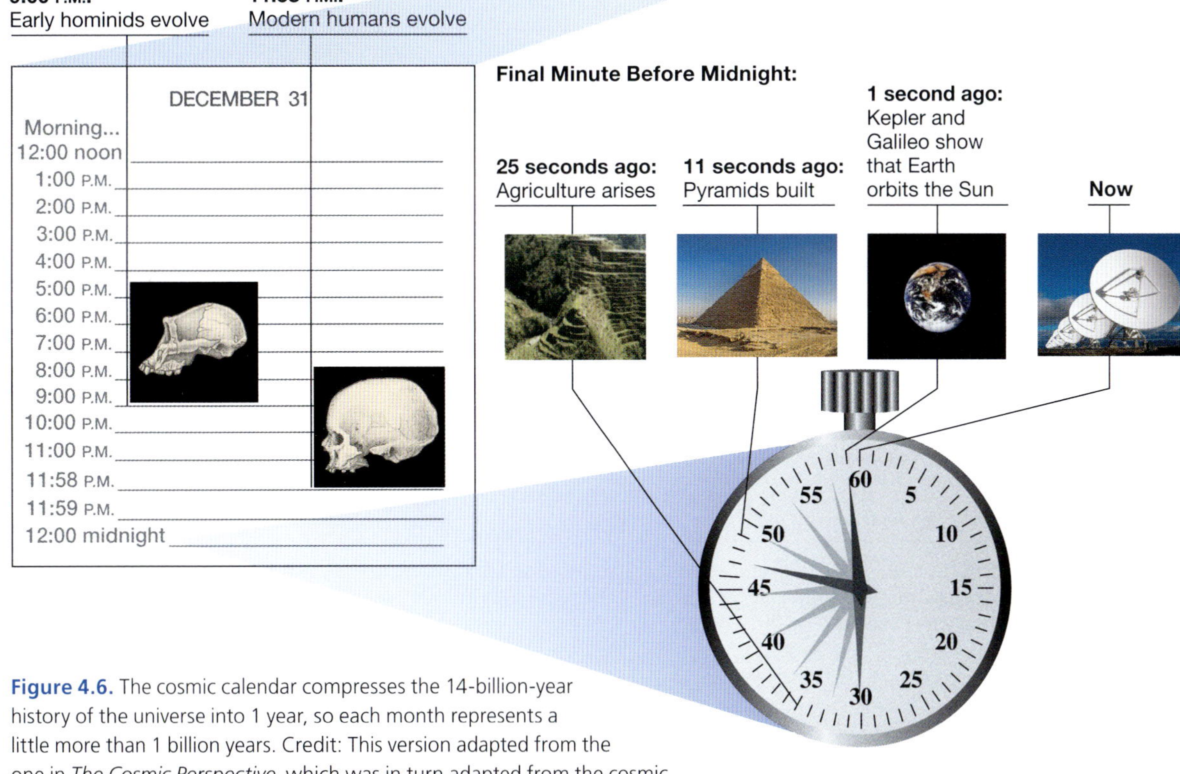

Figure 4.6. The cosmic calendar compresses the 14-billion-year history of the universe into 1 year, so each month represents a little more than 1 billion years. Credit: This version adapted from the one in *The Cosmic Perspective*, which was in turn adapted from the cosmic calendar developed by Carl Sagan (1934–1996).

able animals became prominent only in mid-December. Early dinosaurs appeared on the day after Christmas. Then, in a cosmic instant, the dinosaurs disappeared forever — probably because of the impact of an asteroid or a comet.[TN45] In real time, the death of the dinosaurs occurred some 66 million years ago, but on the cosmic calendar it was only yesterday.

With the dinosaurs gone, furry mammals inherited Earth. Some 60 million years later, or around 9 p.m. on December 31 of the cosmic calendar, early hominids (human ancestors) began to walk upright.

Perhaps the most astonishing thing about the cosmic calendar is that the entire history of human civilization falls into just its last half-minute. The ancient Egyptians built the pyramids only about 11 seconds ago on this scale, and it was only about 1 second ago that we learned conclusively that Earth orbits the Sun rather than vice versa. The average college student was born about 0.05 second ago, around 11:59:59.95 p.m. on the cosmic calendar. On the scale of cosmic time, the human species is a newborn infant, and a human lifetime is a mere blink of an eye.[TN46]

Building a Cosmic Perspective How does the cosmic calendar affect the way you think about human civilization?

What is the significance of the next cosmic year?

Although the cosmic calendar has "now" at the stroke of midnight on December 31, the universe will obviously continue into the next cosmic year (and beyond). For the universe as a whole, nothing much will change as we enter the next cosmic year. Stars will continue to live and die within galaxies, and planets will continue to orbit their stars. Much the same is true for our solar system, since the Sun will live for another 5 billion years — well into May of the next cosmic year.

For us, however, the next cosmic year has great significance. Going even a single second into the next cosmic year represents more than 400 years of real time, and there is no guarantee that our civilization will last that long. We have acquired great knowledge, including our present understanding of how we came to exist, but our technology has also given us multiple ways by which we might bring our civilization crashing down, including nuclear war, global warming, pandemics, and other threats.

So as you contemplate the scale of time illustrated by our cosmic calendar, I hope you'll also think deeply about what it will take for our civilization to continue. We'd like the stroke of midnight that is occurring right now to represent a *beginning* of greater things, not an end. But this will not happen by accident or through luck. It will happen only if all of us work together to build a positive future.

What if we don't make it?

We all hope that our civilization will survive, but the cosmic calendar also offers some perspective on what happens if it doesn't. Remember that it was only *yesterday* on the cosmic calendar that the dinosaurs went extinct. We might therefore expect that in the event of our own demise, it might only be tomorrow (on the cosmic calendar scale) before another intelligent species arises and creates their own civilization. And if they also meet their demise, there could be another civilization on the next cosmic day, and so on.

In real time, this idea translates to the fact that even if it takes tens of millions of years for a new civilization-building species to emerge after an earlier one goes extinct, Earth could have many dozens of future opportunities to become home to a species and civilization that does not destroy itself. Such a civilization might eventually develop the technology to travel to the stars, which means that it could move on and survive even after the Sun dies. If we act wisely, it is still possible that this long-lasting civilization will be ours.

Building a Cosmic Perspective Imagine that the asteroid or comet that killed the dinosaurs had missed, so that dinosaurs did not go extinct. Do you think that dinosaurs would eventually have developed intelligence like ours? Why or why not?

What does this mean for aliens?

Science fiction shows and movies (for example, the *Star Wars* and *Star Trek* franchises) commonly show alien species from different home worlds all sharing similar levels of technology. But when you think about time as illustrated on our cosmic calendar, you'll realize that this is quite far-fetched. After all, on the cosmic calendar we've had our current level of technology for only a fraction of a second, even though our planet has been around for almost four full months. It seems highly unlikely that any other civilization in our galaxy would happen to have reached this same technological level in the same fraction of a second.

If other worlds really have been home to alien civilizations, it's far more likely that such civilizations would be many seconds, minutes, hours, days, or weeks ahead of or behind us on the scale of the cosmic calendar. If they are behind us, then in real time they are still thousands to millions of years away from achieving technology like ours. And if they are ahead of us, they are likely ahead by at least thousands to millions of years.

Adding to this idea, current understanding of the universe suggests that a planet like Earth could have formed as far back as in February on the cosmic calendar, as opposed to the September formation of our planet. In that case, there might be civilizations that arose *months* ago on the cosmic calendar — which means that if such civilizations have survived, in real time they would be *billions* of years ahead of us.

This brings us to a remarkable conclusion. If advanced alien civilizations really exist, then they have almost certainly surpassed our current technological level by thousands, millions, or billions of years. Put a little differently, if our galaxy is currently home to other advanced civilizations, we are almost certainly the most backward of them all. If we ever make contact with such a civilization, we'll have a lot to learn.

Building a Cosmic Perspective Suppose that we someday make contact with an advanced civilization. How do you think it would affect us to know that such a civilization exists?

Are there other ways to put time into perspective?

The cosmic calendar is my personal favorite way of thinking about the scale of time, but of course there are other ways. Here are three others that you might find of interest:

A timeline: You could make a timeline to represent the 14 billion years from the Big Bang to the present. For example, imagine making a timeline that stretches across a classroom. If the classroom is 10 meters (33 feet) long, then each meter represents about 1.4 billion years of time, each centimeter represents about 14 million years, and each millimeter represents 1.4 million years.

A book: Imagine a 1,000-page novel representing the story of the universe. In that case, each page would represent about 14 million years of the 14-billion-year history of the universe. A typical book has about 300 words per page, which means that each word would represent about 47,000 years. So the story of human civilization would not even get a full single word at the end of the book.

The Clock Tower Project: This project has produced a beautiful video that represents time with a DNA-like helix that starts from the Big Bang and gradually unwinds to zoom in on our present moment and possible futures. You can watch it at bigkidscience.com/clocktower.

What does all this mean about *our* place in time?

Much as we found for the scale of space, our *physical* significance is incredibly small in the scale of time. Our entire civilization has risen up in only the last few seconds of the cosmic calendar, and if we wipe ourselves out, the universe as a whole will continue on as if we'd never even existed.

But as I emphasized at the end of Chapter 3, the important part of the story lies in what we have achieved in spite of our physical smallness. We have existed only for a cosmic instant, but in that instant we have figured out much about the long history of the universe. Whether or not other civilizations have done the same, the major message is clear: After some 14 billion years of cosmic evolution, we now represent self-awareness for the universe.

The only question is whether we build on or squander the promise embodied in this 14-billion-year gift from the universe. When you look at the many problems that we face in our world, it's easy to think that squandering is the more likely outcome. Perhaps knowing how we fit into the scale of time might make us a little more likely to avoid that outcome — and to instead put our bickering, hatred, intolerance, and wars behind us so that we might fulfill the promise that awaits us at the dawn of a new cosmic year.

Building a Cosmic Perspective What do *you* plan to do to help ensure that our civilization survives into a new cosmic year?[**TN47**]

Figure 5.1. This photo shows our "spaceship Earth" in true color, made from a composite of satellite-based images. Credit: NASA/GSFC/Reto Stöckli/Visible Earth.

5

Our Future on Spaceship Earth

It is often said that we are all travelers on "spaceship Earth" (**Figure 5.1**). The metaphor comes about from the fact that we are dependent on the ecological systems of our planet in much the same way that astronauts rely on spacecraft life support systems when traveling in space. Moreover, our spaceship Earth really is carrying us on a journey through space, since we travel with it as we rotate daily around its axis, orbit with it around the Sun, travel along with the rest of our solar system around the center of the Milky Way Galaxy, and travel with our galaxy in our expanding universe.

The metaphor should seem even more apt now that you have an understanding of how tiny our planet is in the scheme of the universe. To recap a few of the most important scale ideas that we've covered:

- On our 1 to 10 billion scale, which makes the Sun about the size of a large grapefruit, Earth is smaller than the ball point in a pen and located about 15 meters (49 feet) away from the Sun.
- On that same scale,
    - the Moon — the farthest a human being has ever traveled — lies only about 4 centimeters (1½ inches) away from us;
    - the planets of our solar system are much farther, though you can still walk to the outermost planets in only 10 minutes or so;
    - but to reach the nearest stars, you'd have to walk the distance across the United States.
- To get a sense of the immensity of our Milky Way Galaxy, think about its stars being separated like grapefruits across the United

States at the same time you realize that it contains so many stars that it would take thousands of years just to count them.

- Our galaxy, in turn, is only one of more than 100 billion galaxies in our observable universe. The total number of stars in all these galaxies — and probably also the total number of "Earth-ish" planets (planets that might possibly have surface water and life) — is similar to the total number of grains of sand on all the beaches on Earth combined.
- If we imagine the history of the universe compressed to a single year, our species has existed for only a couple of minutes and our modern technological civilization for only a fraction of a second.

We've discussed some of the key ways in which these ideas ought to change our perspective about ourselves and our planet. In particular, they make clear that we are not the center of the universe, and I've argued that a true understanding of this fact would help us all treat each other with greater empathy, kindness, and respect.

The kind of "cosmic perspective" that we gain from understanding our place in space and time should also teach us to treat our spaceship Earth with much greater care. After all, when we damage the ecological systems on which we depend, we are essentially damaging our own life support systems.

This brings me to one final message. Thousands of past human generations have walked this Earth, but only in the past few generations have we begun to learn how we truly fit into the scale of space and time. The science and technology that has enabled us to learn these things is incredibly powerful, but it is a double-edged sword. On the positive side, it has brought us far greater prosperity than our ancestors could likely have imagined, along with the promise of someday being able to travel throughout our solar system and beyond. On the negative side, it has allowed us to build ever more powerful weapons of destruction and to cause great damage to the life support systems of our spaceship Earth. This damage ranges from direct impacts on ecosystems (such as cutting down forests and driving species to extinction) to the great threat posed by global warming — which, if unchecked, will likely make it impossible for us to maintain the prosperity we have gained.

We therefore find ourselves at what might be the most important turning point in the history of the human species. If we fail to learn the lessons embodied in our cosmic perspective, we may plunge our civilization back into a dark age from which we may never recover. But if we act with forethought and wisdom, we can create social, cultural, economic, and political structures that will allow us to thrive on our spaceship Earth — and ultimately to develop technology that will allow us to travel beyond it.

Imagine for a moment the grand view, gazing across the centuries and millennia from this moment forward. Picture our descendants living among the stars, having created or joined a great galactic civilization. They will have the privilege of experiencing ideas, worlds, and discoveries far beyond our wildest imagination. Perhaps, in their history lessons, they will learn of our generation — the generation that history placed at the turning point and that managed to steer its way past the dangers of self-destruction and onto the path to the stars.

Teacher Notes

The following Teacher Notes (referenced in the main text by "**TN**" and a number) are designed especially to help teachers but may also be of interest to others. Some of them provide additional background that will be helpful when students ask questions; others suggest activities that you may wish to do with students.

TN1 More technically, the colors in the JWST images are assigned based on the infrared wavelengths that the telescope cameras record. Just as longer wavelength visible light is red and shorter wavelength visible light is blue, the images are shown by assigning red to the longer infrared wavelengths and blue to the shorter infrared wavelengths (and other colors of the rainbow to wavelengths in between). Although this choice of colors is artificial since our eyes cannot see these wavelengths at all, it is very useful in at least two ways. First, the colors tell us something about the wavelengths of light recorded in the images. Second, the images end up looking familiar in the sense of appearing similar to what we see in many visible-light images, which makes it easier for us to understand and interpret the images. If you would like to explore this particular photo (Figure 1.1) in more detail, you can find a zoomable version by searching "SMACS 0723" (the name of the main galaxy cluster seen in the image) on the JWST website (webbtelescope.org).

TN2 The spikes that appear in astronomical photographs like Figure 1.1 occur because of the way light interacts with the arrangement of the telescope's mirrors and the support beams that hold them in place. The technical name for this process is *diffraction*, so the spikes are often called "diffraction spikes." The spikes are noticeable only around single bright points of light, such as relatively bright stars. This makes it easy to distinguish between foreground stars and distant galaxies in images like Figure 1.1, since only the foreground stars have spikes.

Also worth noting in Figure 1.1: Although many of the galaxies appear to be shaped like small arcs, the galaxies don't really have these arc shapes. Instead, the arcs are caused by something called *gravitational lensing*, in which light is magnified or bent by the gravity of a massive object. Gravitational lensing is an important prediction of Einstein's *general theory of relativity* that has been verified repeatedly by images like that in Figure 1.1. In this case, the "massive object" is a cluster of galaxies (called SMACS 0723) located about 4.6 billion light-years away; most of the normally shaped galaxies in Figure 1.1 are part of this cluster. The strong gravity of this cluster then distorts the light from more distant galaxies, causing the arc shapes. Moreover, the magnification caused by the cluster's gravitational lensing allows us to see galaxies that would otherwise be too dim for the telescope to record, making it a very useful effect. Indeed, this magnification is a major reason why astronomers chose this cluster as one of the first observing targets with JWST, since they knew that its gravitational lensing would enable them to see some of the most distant galaxies in the universe.

TN3 A technical note on large distances: Because the universe is expanding (as discussed in Chapter 4), there is more than one way to describe distances to far-away galaxies. I've chosen to state distances based on how much time it has taken for an object's light to reach us, which is what astronomers call the object's "lookback time." For example, when I say that a galaxy is 1 billion light-years away, I mean that its light has taken 1 billion years to reach us — which means that we are seeing it as it looked 1 billion years ago. Similarly, when I say that a galaxy is 13 billion light-years away, I mean that its light has taken 13 billion years to reach us — so we are seeing it as it looked 13 billion years ago (which is less than 1 billion years after the universe was born). You may occasionally see distances described by a different method, but in my opinion the method I've chosen is by far the simplest and clearest.

One other terminology note: Because some of the most distant objects in the JWST images look like small red smudges, astronomers refer to them as "little red dots." **Figure TN3.1** shows a close-up of one of these; scientists suspect that they may represent glowing gas around black holes in very young galaxies.

Figure TN3.1. This image is a close-up of a very distant object (more than 13 billion light-years away) from a larger JWST image (similar to the one in Figure 1.1). You can see why such objects are sometimes referred to as "little red dots." The object's name is at the top, and z represents its measured redshift, which astronomers can use to calculate distance. Credit: NASA, ESA, CSA, STScI, D. Kocevski (Colby College).

TN4 If you'd like to calculate the distance represented by a light-year for yourself, the starting point is knowing the basic relationship among distance, speed, and time:

$$\text{distance} = \text{speed} \times \text{time}$$

This formula is easy to remember with a simple example such as "How far does a car go if you drive at a speed of 50 kilometers per hour for 2 hours?" The answer is 100 kilometers, because you will travel 50 kilometers in the first hour and another 50 kilometers in the second hour — which is the same answer you get by multiplying the speed (50 km/hr) by the time (2 hr).

The same idea allows us to calculate the distance represented by a light-year simply by multiplying the speed at which light travels (the speed of light, or 300,000 km/s) by the time for which it is traveling (which is 1 year if we are calculating a light-year):

$$1 \text{ light-year} = \text{speed of light} \times 1 \text{ year}$$

The only trick is that because the speed of light is given as 300,000 kilometers per second, we must do this multiplication while also converting a year into seconds. You can do this easily with a calculator: Start by putting in 300,000 for the speed of light in kilometers per second, then multiply by 60 (for the 60 seconds in 1 minute), multiply by 60 again (for the 60 minutes in 1 hour), multiply by 24 (for the 24 hours in 1 day), and multiply by 365 (for the 365 days in 1 year). The full calculation with units looks like this:

$$\begin{aligned} 1 \text{ light year} &= (\text{speed of light}) \times (1 \text{ yr}) \\ &= \left(300{,}000 \frac{\text{km}}{\text{s}}\right) \times \left(1 \cancel{\text{yr}} \times \frac{365 \cancel{\text{days}}}{1 \cancel{\text{yr}}} \times \frac{24 \cancel{\text{hr}}}{1 \cancel{\text{day}}} \times \frac{60 \cancel{\text{min}}}{1 \cancel{\text{hr}}} \times \frac{60 \text{ s}}{1 \cancel{\text{min}}}\right) \\ &= 9{,}460{,}000{,}000{,}000 \text{ km} \end{aligned}$$

You can confirm with a calculator the final result shown above, which tells us that a light-year is about 9.46 trillion kilometers (5.88 trillion miles). (Recall that 1 trillion is a 1 followed by 12 zeros.) To make this distance easier to remember, in the main text I've rounded it up to "about 10 trillion kilometers" (6 trillion miles).

TN5 You may occasionally hear large distances described in terms of *parsecs* rather than light-years; 1 parsec is equal to about 3.26 light-years.

In case you are wondering, the term *parsec* arises from the major method that astronomers use to measure the distances of nearby stars, which is called *stellar parallax*. **Figure TN5.1** shows how stellar parallax works. As Earth orbits the Sun each year, a nearby star will appear to slightly shift its position relative to more distant stars. Notice that the geometry forms a right triangle in which we know the base, which is the Earth-Sun distance of about 150 million kilometers (93 million miles), or 1 AU. Therefore, by measuring the parallax angle, *p*, we can use trigonometry to calculate the distance to the nearby star. (This is essentially the same technique as the triangulation used

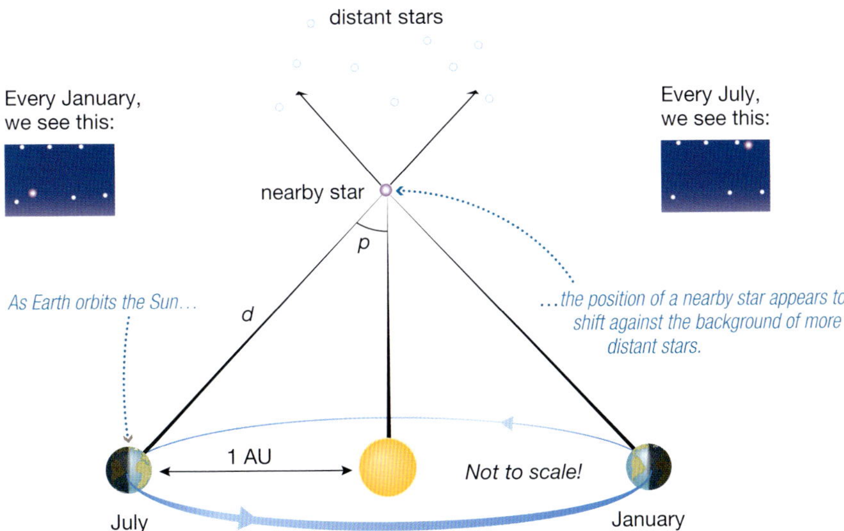

Figure TN5.1. Stellar parallax is the apparent shift in the position of a nearby star relative to more distant stars that occurs as we orbit the Sun. The diagram is not to scale, since the actual distance to even the nearest star is about 270,000 times the Earth-Sun distance (270,000 AU, or almost 4.3 light-years). Credit: *The Cosmic Perspective*.

in surveying.) The figure is not to scale, and in reality the angle *p* is always smaller than 1 arcsecond. (Recall that there are 360 degrees in a circle, each degree can be subdivided into 60 arcminutes, and each arcminute can be subdivided into 60 arcseconds.)

The term *parsec* is an abbreviation for *parallax second* and represents the distance of a hypothetical star with a parallax angle of 1 arcsecond, which a trigonometric calculation reveals to be 206,265 AU, or about 3.26 light-years. Note that an arcsecond is an extremely small angle; in fact, it is *less than the angular width of a human hair viewed across the length of a football field.* Nevertheless, stars are so far away that even the nearest have parallax angles smaller than 1 arcsecond. This means that we cannot notice stellar parallax with the naked eye. Instead, parallax angles are measured by comparing telescope observations at different times of year.

Interestingly, the fact that stellar parallax cannot be detected with the naked eye was used by scholars in ancient times to argue *against* the idea of Earth going around the Sun. The reason is that, as Figure TN5.1 shows, stellar parallax is an inevitable consequence of an orbiting Earth. The ancient scholars' inability to detect this parallax therefore made most of them think that Earth must be stationary. For much the same reason, once parallax was finally detected with the aid of telescopes, it represented direct proof that Earth really is a planet orbiting the Sun.

Note: Ancient scholars recognized that an alternative explanation for their inability to detect parallax was that the stars were too far away for it to be noticed by eye. However, they could calculate how far away this would make the stars, and in most cases they thought it inconceivable that the stars could be so far away, even though it turns out that they are.

TN6 The idea that looking out in space means looking back in time often leads students (and others) to ask, "If we can only see stars as they were in the past, how do we know they are still there?" The answer relies on the fact that our studies of the universe have given us an understanding of the time scales that are involved in major changes. For stars, major changes generally occur quite slowly, over millions or billions of years. Therefore, if we look at stars that are "only" hundreds or even thousands of light-years away, most will not have changed in any significant way since their light started its journey toward us.

However, there are a few cases in which major changes might have occurred, and at least two are worth noting. The first is for stars that are near the ends of their lives, when they can in some cases die in spectacular explosions that we call *supernovae*. As an example, the bright star Betelgeuse (the reddish "shoulder" star in the constellation Orion) is thought to be close to the time when it will end its life in a supernova — though "close" in this case means the supernova might occur any time within about the next 100,000 years. Still, because Betelgeuse is located roughly 700 light-years from Earth, it is possible that Betelgeuse has already exploded, since light from the explosion would not yet have reached us if this death occurred within the past 700 years.

The second case is for galaxies that are extremely far away. Galaxies undergo dramatic changes over hundreds of millions to billions of years. Some of these changes are simply a result of the births and deaths of the stars within them, while others may be due to such things as collisions or mergers with other galaxies. Therefore, when we look at galaxies that are hundreds of millions to billions of light-years away, they probably look quite different than they would if we could see them as they look today.

TN7 Current technology only allows us to search for exoplanets around stars that are relatively nearby, which means that all the statistics about exoplanets are based on stars within our Milky Way Galaxy. However, we can safely assume that similar statistics would apply to stars in most other galaxies as well. We can be confident in this assumption because we know that stars throughout the universe are made of the same chemical elements and obey the same physical laws, including the law of gravity. (We can determine chemical compositions even of very distant stars and galaxies through spectroscopy, and we can verify that they obey the same physical laws by measuring such things as orbital speeds.) This implies that stars and planets should form through the same basic processes in all galaxies, which in turn implies that the numbers of exoplanets should follow the same general statistical patterns in all galaxies.

TN8 Some astronomers suspect that the far outer solar system may harbor an as-yet-undiscovered "Planet 9" that may be 5 to 10 times as massive as Earth. The reason for this suspicion comes from analysis of the orbits of objects found beyond Neptune (in the Kuiper belt). The dwarf planets Pluto and Eris are the most massive of these objects, but thousands of smaller ones have also been detected. Careful study suggests that the orbits of these objects may be clustered in a way that could be explained by the gravitational influence of an unseen Planet 9. However, some astronomers suggest the clustering may not even be real (instead appearing as a result of bias in current observational methods) or may have a different cause.

If it really exists, Planet 9 would be so far away (perhaps 15 to 30 times as far as Neptune) that it would be extremely dim, which means finding it would require searching the sky with some of our most powerful telescopes. Such a search should be hitting high gear around the time this book is published, thanks to the 2025 opening of the Vera C. Rubin Observatory (**Figure TN8.1**). This observatory, located on a mountain called Cerro Pachón in Chile, is undertaking the most comprehensive sky survey to date. Its telescope is unique in both being very large (8.4-meter-diameter primary mirror) and having a very wide field of view (each image it takes will cover an angular area equivalent to that of 50 full moons). As a result, it can obtain images of the entire sky visible from its mountaintop every few nights. Comparisons of such images (in essence, making a time-lapse movie of the sky) over the expected 10-year length of the study are expected to lead to many discoveries.

Figure TN8.1. The Vera C. Rubin Observatory, seen here under the Milky Way. Credit: Rubin Observatory/NSF/AURA/B. Quint.

And, unless the claimed Planet 9 is much dimmer than would be expected, the Rubin Observatory ought to allow us either to find it or to rule out its existence within the next decade or so.

TN9 You may wonder why the discovery of a moon (Charon) around Pluto was important to learning Pluto's mass and size. The answer is that when we discover a new object in the sky, we are really just seeing a dot or patch of light, and that alone does not tell us much about the object's nature. We can often learn a lot by studying the light in detail (for example, we can learn chemical composition and much more through spectroscopy), but we usually need other information to measure an object's size and mass.

Let's start with size. In principle, we can calculate an object's true size from its angular size in the sky if we also know its distance. In the case of Pluto, astronomers knew its distance from its observed orbit around the Sun, but early measurements of its angular size were not precise enough to allow an accurate size determination. The discovery of Charon changed this because Charon's orbit sometimes has it passing in front of or behind Pluto, and the timing of these "occultations" allowed much more precise determination of the sizes of both Pluto and Charon. Turning to mass: The only sure way to determine an object's mass is by observing its gravitational effects on some other object orbiting it or passing it by. The discovery of Charon gave us such an orbiting object. More specifically, Charon's orbital characteristics made it possible to calculate Pluto's mass from the law of gravity (through an equation known as Newton's version of Kepler's Third Law).

To sum up: Prior to the 1978 discovery of Charon, astronomers already had strong hints that Pluto was much smaller and less massive than any of the other planets. But Charon's discovery allowed precise determination of Pluto's size, mass, and density, thereby confirming that Pluto differed from the other planets in fundamental ways.

TN10 Here is a little more detail about the official definition of a planet and why the number of dwarf planets is not yet known. Official decisions on astronomical names and definitions rest with the International Astronomical Union (IAU), an organization of professional astronomers from around the world. In 2006, the IAU adopted a definition essentially stating that a *planet* is an object that (1) orbits a star, (2) is large enough for its own gravity to make it round (see "Why are planets round?" on page 14), and (3) has cleared most other objects from its orbital path. The idea behind the third criterion is that a planet should be gravitationally dominant in its orbit, meaning that its gravity would essentially have swept up any smaller objects lying near its orbital path.

A *dwarf planet* was defined to be an object that meets the first two criteria but not the third. Pluto and Eris fall into this category because although they meet the first two criteria (they orbit the Sun and are round), they don't meet the third because they share their region of the solar system — the Kuiper belt — with a vast number of other objects. Astronomers estimate that the Kuiper belt contains at least 100,000 objects that are larger than 100 kilometers (60 miles) across (and millions of smaller objects).

The reason the exact number of dwarf planets is unknown is that it is very difficult to determine the masses and shapes of these distant objects. It's likely that at least several dozen (and perhaps several hundred) objects in the Kuiper belt are round and hence would qualify as dwarf planets, but in most cases current technology does not yet allow us to know for sure whether they are indeed round.

Perhaps also worth noting: Those astronomers who opposed the 2006 IAU definition generally argued in favor of defining a planet only by the first two criteria, in which case there would currently be 13 planets in our solar system (the eight planets plus the five confirmed dwarf planets), with the total number still unknown. My own opinion is that the 2006 definition was the right call, primarily because I think it is easier to teach kids about the solar system if we can say there are eight planets plus lots of smaller objects (including dwarf planets) as opposed to explaining the subtleties of a definition that would have left the number of planets unknown.

TN11 A little more detail on the history of definitions of the word *planet*: The "about 400 years ago" (when the list of planets was first considered to include Earth) refers to the Copernican revolution, which is generally considered to have begun in 1543 with the publication of a book by Nicolaus Copernicus claiming that Earth orbits the Sun. (The idea of Earth orbiting the Sun was also suggested in ancient times, but few people took it seriously until after Copernicus's book came out.) Evidence supporting this claim was solidified by about 1610 through the work of Johannes Kepler and Galileo, though it didn't gain a firm theoretical foundation until Isaac Newton published the law of gravitation in 1687. Although there was no formal body in charge of astronomical definitions back then, it became widely accepted that a "planet" was an object that orbited the Sun.

This new definition worked well for a while. In particular, it made it easy to add Uranus and Neptune to the list of planets after their discoveries in 1781 and 1846, respectively. But things got complicated as astronomers began to identify many smaller objects — what we now call *asteroids* (a few of which were discovered before Neptune) — orbiting the Sun. At first these objects were also called planets (or "minor planets"), but as their numbers grew, scientists gave them their own category.

Interestingly, even before the discovery of Uranus, astronomers already realized that comets orbit the Sun. But it does not appear that anyone thought of them as planets, perhaps because they look so different from planets in the sky and because their orbits are so elliptical (stretched out) compared to the nearly circular orbits of the planets.

It's worth noting that the current IAU definition of *planet* also presents at least some difficulties. For example, astronomers have now identified many "free-floating planets," meaning planet-size objects found in interstellar space. The leading hypothesis for these objects is that they were once planets orbiting a star but were subsequently kicked out of their orbits by gravitational interactions. But should they still be called planets when they no longer orbit a star? Consider also the possible Planet 9: Suppose we learn that it really exists and really is 5 to 10 times as massive as Earth but that it shares its orbital region with many smaller objects. The current IAU definition would then make it a dwarf planet (because it would not meet criterion 3 of the planet definition), even though it would be much larger than Earth. And even if Planet 9 turns out not to exist, it's very likely that we'll discover objects with such properties orbiting other stars, forcing us to confront the same definitional issue.

As with the case of Pluto, the take-away message should be that nature does not always accommodate our human desire for clear categories. The definition of *planet* has long been somewhat fuzzy, and it is possible that we may never come up with a perfectly clear definition of the term.

TN12 Humans have known that Earth is round for at least 2,500 years, as this idea was already being taught by ancient Greek philosophers (notably by the famous mathematician Pythagoras) before 500 B.C.E. Nevertheless, even today you can still find a few people insisting that Earth is flat. Because these "flat Earthers" are often quite vocal (especially on social media and the internet), it's likely that you'll have friends or family members — or students, if you are a teacher — who will wonder how we *know* that Earth is round. In fact, there are many ways to prove beyond any doubt that Earth really is round. Here I'll give you a few of the major ones.

The most obvious way we know that Earth is round is simply by looking at photos or videos of Earth from space. (For example, see Figure 5.1 in this book or the great time-lapse showing a full year of Earth's rotation posted at bigkidscience.com/Earthrotation.) Of course, ancient people figured out that Earth was round long before we had such views from space. They did this based on at least three key observations:

1. They noticed that the visible constellations change as you travel north or south, which means at a minimum that Earth must be curved in those directions.
2. They noticed that when ships sailed into view on the horizon their masts gradually appeared as they approached, and they noticed that the horizon became farther away when viewed from higher elevations on hilltops or mountains. These observations prove that there are parts of the world that are hidden below the horizon no matter which direction you look, which means Earth must be curved in all directions.
3. They noticed the curved shadow of Earth on the Moon during lunar eclipses. If you experiment with shadows from a sphere and a coin, you'll quickly see that the coin gives a round shadow only in a few particular orientations. The fact that the shadow of Earth during eclipses is *always* curved therefore means that Earth must be spherical, because only a sphere will always cast a round shadow, regardless of orientation.

These observations, all made easily in ancient times, leave no doubt that Earth is round. But for people today who might have questions, there are many more observations that anyone can make easily for themselves. For example:

- *Airplane flight:* Anyone who has flown on an airplane has seen the horizon distance increasing with altitude, which can only occur if Earth's surface is curved. The fact that the same thing happens anywhere in the world means the curvature must wrap all the way around, proving that Earth is round.
- *Clouds:* On a day with scattered clouds, all the clouds will be at high altitudes above the ground. But as you look toward the horizon, you will notice that the clouds appear to be lower and lower until they intersect the horizon in the distance. This could not happen if Earth were flat, because in that case a horizontal line of sight would always remain the same distance above the ground.
- *Daytime and nighttime:* Call or text someone who lives many time zones away from you. For example, if you live in the United States, call someone in Australia. You'll find that if it is daytime for you, it is nighttime for your friend (or vice versa). If Earth were flat, the Sun angle would be about the same from all locations, so it would not be possible for it to be daylight in some places and night in others. (To prove the north-south curvature, you can use changes in the Sun's exact path across the sky or changes in the orientation of constellations and the Moon that occur with changes in latitude.) Note: Some flat Earthers try to explain this away by claiming

that the Sun shines like a spotlight, but if you make a model using a "spotlight Sun," you'll find that it cannot reproduce observed time differences around the world.

- *Shadows in different locations:* Call or text a friend who lives at least a few hundred kilometers away from you (in any direction) while each of you take a ruler of the same length and stand it straight up in a sunny location. Use your cellphones to take pictures. Compare the shadows at your two locations, and you will see that one shadow is longer than the other — which makes sense only on a curved Earth.

This is only the beginning of the list of ways in which you can debunk flat Earth claims; you'll find many more discussed (along with several classroom activities) in Section 3.1.1 of my online *Earth and Space Science* text posted at grade8science.com. For those who'd like to understand some of the psychology behind flat Earth claims, I highly recommend the documentary *Behind the Curve* (2018).

TN13 For a deeper understanding of the shapes of astronomical objects, we must recognize that while the force of gravity *tends* to make objects round (as shown in Figure 2.5), other forces can resist this tendency. The strength of these forces depends largely on the phase (solid, liquid, or gas) of the material in the object.

Let's start with the Sun, which is essentially a giant ball of very hot gas. Because gas tends to move easily, it does not resist the force of gravity that tries to shape it into a sphere. That is why the Sun and other stars are spherical in shape. Next consider the largest planets of our solar system: Jupiter, Saturn, Uranus, and Neptune. These planets also formed largely from gas (though today their interiors are largely in a liquid-like state) and therefore were also easily shaped into spheres by gravity.

At the other end of the size spectrum, it's also easy to understand the shapes of small objects like asteroids and comets. These objects are made mostly of solid rock or solid ice, both of which can resist the compression of gravity. Because their small sizes mean their gravity is relatively weak, the solid resistance can successfully counter gravity's tendency to make objects spherical. That is why these objects can have almost any random "potato shape." (Note: This also relates to the common child question "Do I have gravity?" The answer is yes, since all objects exert gravity, but because humans [and other everyday objects] are quite small, the gravitational forces we exert on ourselves are very weak and cannot overcome the internal forces that give us our bodily shapes.)

The most challenging cases to understand are objects that are solid (or mostly solid) but round. These objects include Earth and the other inner planets, a few large moons (including our Moon), and dwarf planets like Ceres and Pluto. Given that solid material can resist gravity, how did these worlds become round? Two key processes were involved:

1. Many of these worlds, including Earth, became so hot during early stages of their formation that their rock melted, becoming molten (liquid). During this time, gravity was able to shape the molten material into spheres quite easily. This also explains how planets like Earth ended up with dense cores and lower-density mantles and crusts: During the time when their interiors were molten, the denser material sank toward the center while lower-density material rose upward.
2. Even for worlds that never became hot enough to melt, gravity can still shape them into spheres if they are large enough and enough time passes. The reason this can happen is because "solid" material is not perfectly rigid and can change its shape if you give it enough time. You can see this fairly easily by recognizing the gradual flow of the solid ice of glaciers; the same can also happen with rock, though more slowly.

Considering both processes together, calculations tell us that gravity will cause any rocky or icy object larger than about 500 kilometers (300 miles) across to become spherical in less than about 1 billion years. Our solar system is much older than this (about 4½ billion years old), which is why virtually all objects of this size and larger in the solar system are expected to be round.

One more related note: Despite the action of gravity, large objects can deviate from being perfect spheres for at least two reasons. First, if an object rotates relatively fast, the forces associated with rotation will tend to make it bulge at its equator. For example, rotation explains why Earth's equatorial diameter is about 0.3% larger than its polar diameter. Second, although gravity may have compressed an object into a generally spherical shape, solid rock or ice near the surface may not have been fully reshaped, especially if there have been recent impacts or geological activity. That is why we see impact craters, mountains, valleys, and other structures on the surfaces of solid worlds.

TN14 In addition to the listed "objects" that make up the Sun's family, the solar system contains countless particles too small to be considered asteroids and some extremely low density gas. The particles range in size from microscopic bits of interplanetary "dust" to pebbles and small rocks. This material has many noticeable effects. For example, *meteors* (also called "shooting stars") are flashes of light that we see as some of these particles (typically pebble-size) burn up in Earth's atmosphere, and the *zodiacal light* (a faint evening glow that traces out the ecliptic if you look from a place with a very dark sky) is caused by light reflecting off interplanetary dust particles.

TN18 The origin of asteroids and comets will be clearer if we go into a little more detail about the formation of the solar system. The cloud from which our solar system formed was primarily gaseous (with small amounts of dust mixed in). Gravity could form the Sun at the center because, as discussed already, gravity always pulls toward a center. But how did planets form in the disk of gas that surrounded the Sun?

According to current understanding, the formation of the planets began with small, solid particles condensing from the gas in much the same way that solid snowflakes or hailstones condense from water vapor in clouds on Earth. These particles collided with each other frequently, sometimes sticking together and growing larger. As they grew, their gravity strengthened, drawing in more material and allowing some of them to grow into full-fledged planets. **Figure TN18.1** summarizes this basic process for the inner solar system. A similar process is thought to have occurred in the outer solar system, except that those planets grew so large that their gravity enabled them to capture gas, which allowed them to grow further until they became the "giant planets" Jupiter, Saturn, Uranus, and Neptune (sometimes called "gas giants," though their interiors are actually mostly in liquid-like states). (You can find more detail on this process in Section 4.3.1 of the online *Earth and Space Science* text posted at grade8science.com.)

This process also meant that the planets gradually "swept up" most of the remaining material in their orbital vicinities; you may recognize this idea as

Figure TN18.1. This sequence illustrates how the planets in the inner part of our solar system formed. Credit: Adapted from *The Cosmic Perspective*.

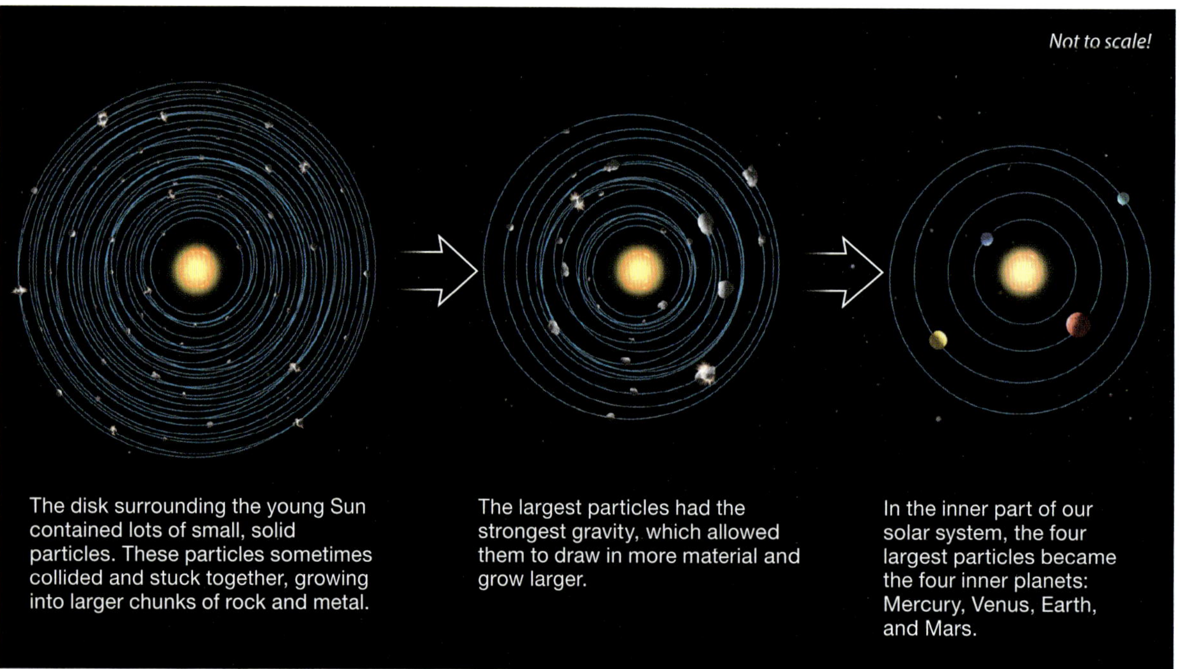

the third criterion of the IAU planet definition we discussed in Teacher Note 9. But the sweeping did not catch everything, so there were inevitably a lot of solid chunks that were never swept up. Over time, some of these "leftovers" crashed into the Sun, planets, or moons (which explains the impact craters that we see on the Moon and other solid worlds), but some remain today as the objects that we call asteroids and comets.

The distinction between asteroids and comets can be traced back to the first step in the formation process, which was condensation of solid material from the cloud:

- In the inner solar system, where temperatures were fairly high, only rocky material (which includes both rock and metal) could condense from the gas to make solid particles. (It was too hot for anything that we might call "ice" to condense from the gas.) Therefore, the solid chunks of the inner solar system were made of rock, explaining the rocky composition of the inner planets and of asteroids.
- Farther from the Sun, where temperatures were much colder, it was also possible for substances such as water (H_2O), methane (CH_4), and ammonia (NH_3) to condense to make solid particles of ice. Therefore, the solid chunks of the outer solar system were ice-rich, explaining the icy compositions of comets and of most outer solar system moons.

You may also wonder why asteroids are concentrated in the asteroid belt and comets in the Kuiper belt and Oort cloud. The asteroid belt is the easiest to understand: It exists in the region between Mars and Jupiter, where the competing effects of gravity from the Sun and gravity from Jupiter allow for stable orbits that are unlikely to ever crash into a planet. In other words, the asteroids of the asteroid belt have been able to survive for the 4½ billion years since the solar system first formed, while most of those that once existed in other parts of the inner solar system crashed into planets (or moons or the Sun) long ago. The explanation for the Kuiper belt is somewhat similar: It is a region beyond Neptune where icy comets that formed long ago can still survive. The origin of the Oort cloud lies with the comets that were left over in regions *between* the four outer planets (Jupiter, Saturn, Uranus, Neptune). There, the "sweeping" done by the gravity of these large planets had two related effects: It caused some objects to crash into one of the planets (or moons or the Sun) while speeding up others so that they were flung into the far outer solar system — where they make up the Oort cloud of today.

TN19 A couple of notes on questions that may arise about the spinning gas cloud from which our solar system formed (called the *solar nebula*): First, while the ice skater analogy explains why the cloud spun faster as it shrunk, you might wonder what gave it any spin in the first place. The answer is that some small net spin was almost inevitable. Any giant cloud in space is made up of countless individual particles (mostly gas atoms and molecules), and as gravity begins to make it contract, every particle has its own orbit around the center of the cloud. Although these initial orbits would be essentially random, it would take an extreme coincidence for every orbit to be exactly matched by an opposite orbit of another particle. As a result, there is almost always some small net motion in some direction around the center; this represents the initial rotation of the cloud. This rotation would be imperceptible at first, but it is magnified as the cloud shrinks in the same way that the skater's rotation is magnified as he brings in his arms. (For those who have studied physics, this process represents the law of conservation of angular momentum in action.)

Second, while the pizza dough analogy is useful in explaining why a cloud flattens into a disk, it may raise questions about why the solar nebula shrank as it flattened instead of spreading wider like a spinning pizza dough. The answer is that we need to distinguish the flattening part from the shrinking/spreading part. The flattening occurs for a similar reason in both cases: Once there is a preferred axis of rotation (as there is in both cases due to the increasing spin rate), the general motion will force all the particles into a flattened disk rotating around the axis. The solar nebula keeps shrinking as this occurs because of the action of gravity. The pizza dough spreads out because there is no force holding it in, so the centrifugal force of rotation pushes the spinning dough outward.

TN20 A small number of exceptions to this general trend (in which planets orbit their star in the same direction and approximately the same plane) have been observed, but they are readily explained. In particular, gravitational interactions among forming planets can cause significant orbital changes. In rare cases, this can result in a planet orbiting at a significant inclination to the general orbital plane (or even going backward). Note that this is more likely to happen with less massive objects (because it is easier to change their orbits), which is why we see so many smaller objects like asteroids and comets with inclined orbits even here in our own solar system.

TN21 As discussed earlier, I am describing distances based on light travel times (what astronomers call lookback times). Therefore, because light from the edge — or "horizon" — of the observable universe has taken 14 billion years to reach us, I say that this horizon is 14 billion light-years away. You may occasionally find other sources (books, websites, etc.) that factor in the effects of the universe's expansion on distances, in which case they might state a much larger distance for the horizon (such as 46 billion light-years). This is simply an alternative (and, in my opinion, inferior) method of describing large distances and does not represent any difference in scientific understanding.

To be a little more specific: The fact that the universe is expanding means that the horizon must have been much closer than 14 billion light-years away when the light we now see from it started its journey and must be much more than 14 billion light-years away today. It is possible to use the expansion rate to calculate what today's distance must be, and the result is that the horizon must now be some 46 billion light-years away. Again, while such an approach is legitimate, I personally find it confusing and somewhat pointless, which is why I will stick with describing distances based directly on light travel times.

TN22 Many students are likely to have heard of the multiverse, usually because they've come across the idea in some science fiction book or show. It's good to encourage interest in such speculative ideas, but it's also important to distinguish speculation from evidence-based science. Moreover, because the multiverse is speculative, there are many different versions of the idea. Personally, I'm skeptical of the idea that a multiverse actually exists — but since this is purely an opinion, I could well turn out to be wrong. Note: If you are teaching high school or college students, you might consider a discussion of whether the concept of a multiverse is testable and, if not, whether it should qualify as "science."

TN23 If you are a classroom teacher, I recommend formalizing the Earth-Moon scale activity by pairing up students, with one student holding the model Earth and the other holding the model Moon. Each pair can then make a guess at the correct placement of the Moon, and you can compare and discuss the various guesses by your class. Note: I have posted a more formal version of this activity (using a slightly different scale that will be easier to use in a classroom), which you can find by going to grade8science.com and selecting Section 1.2.2.

TN24 The astronaut quotes that I've included offer examples of what is sometimes called *the overview effect* (a term coined by author Frank White), defined as the perspective shift that many astronauts report upon seeing Earth from space. A simple activity for students is to have them search for other great astronaut quotes about what it means to see our planet from space. You might then work as a class to compile a list of favorite quotes, perhaps taking time to discuss them in greater depth.

TN25 Many students will wonder how many Earths could fit *inside* the Sun. To find this answer, we need to consider a *volume* comparison rather than a diameter comparison. You probably know that the volume of a cube is its side length to the third power. For example, if a cube is 10 centimeters on a side, its volume is $10^3 = 1{,}000$ cubic centimeters. In fact, all volumes scale as the third power, or cube, of length. This fact gives us an easy way to figure out about how many Earths could fit inside the Sun. Because the Sun's diameter is about 109 times that of Earth, its volume is about $109^3 \approx 1{,}300{,}000$ times Earth's volume. In other words, if we could stuff them in without any wasted space (that is, cutting some of them up to fill in gaps between spheres), *about 1.3 million* Earths would fit inside the Sun!

TN26 Solar "activity" is a general term referring to such things as sunspots and solar flares that can have impacts on Earth and other planets. Solar activity goes up and down in an approximately 11-year cycle, sometimes called the *solar cycle* (or *sunspot cycle*). Therefore, the Sun is always getting more active about half the time — and less active the other half. For example, the Sun reached its most recent maximum activity around the time I completed this book in 2025, which means it is expected to become less active over the following few years. In other words, the "angry Sun" claim was pure coincidence and simply reflected the fact that we were in an upswing of the solar cycle at the time. A more general answer to this Building a Cosmic Perspective question: The Sun is so enormous in comparison to Earth that nothing that happens here on Earth (whether or not it is caused by humans) can have any significant effect on the Sun. (Note: The *Time* cover with the "angry Sun" claim was the July 3, 1989 issue.)

TN27 In creating scale models of the solar system, there is only a fairly narrow range of scales over which the planets are big enough to be seen while the length of the exhibit is short enough to be easily traversed by children, the elderly, and people with disabilities. The Voyage team (of which I'm a member) chose the 1 to 10 billion scale because (1) it is within that narrow range of scales; (2) the factor of 10 billion makes calculating scaled values fairly easy even for upper elementary and middle school students (see Teacher Note 28); and (3) a light-year becomes about 1,000 kilometers on this scale (see Teacher Note 32), so it is easy to start thinking about distances to the stars. Note: As mentioned in the caption to Figure 3.8, visit voyagesolarsystem.org for a list of existing Voyage models or to learn how to get a Voyage model for your own campus or community. (You may also wish to see the info posted on my personal website [jeffreybennett.com] under the "Model Solar Systems" tab, and you can feel free to email me for more information about obtaining a Voyage model.)

TN28 Voyage models include the dwarf planet Pluto primarily because it is so well known (and because the first models were installed before Pluto lost its status as an official planet). The models do not include any other dwarf planets. However, they include a pedestal with information about asteroids and comets that is located at the scaled distance of the dwarf planet (and largest asteroid) Ceres.

TN29 You (or your students) can calculate sizes and distances on the 1 to 10 billion Voyage scale simply by looking up the actual values and dividing them by 10 billion. Just be sure you do it with metric units (otherwise the unit conversions are far more difficult). The only trick is that since you'll generally find actual values given in kilometers, you'll need to do some unit conversions to make the scaled values meaningful. The following three examples show the idea. (Note: The calculations below show the numbers written out fully, but the process is even easier if you are familiar with and use scientific notation.)

Example 1 — Size of the Sun. You can look up the fact that the Sun's real diameter is about 1,400,000 kilometers. We therefore find the diameter of the model Sun by dividing its real diameter by the scale factor of 10 billion (10,000,000,000):

$$\frac{1{,}400{,}000 \text{ km}}{10{,}000{,}000{,}000} = 0.00014 \text{ km}$$

We can make this answer more meaningful by converting it to centimeters, which we can do by using the facts that there are 1,000 meters in a kilometer and 100 centimeters in a meter:

$$0.00014 \text{ km} \times \frac{1{,}000 \text{ m}}{1 \text{ km}} \times \frac{100 \text{ cm}}{1 \text{ m}} = 14 \text{ cm}$$

The Sun on the Voyage scale is about 14 centimeters (5.5 inches) in diameter, which is about the size of a large grapefruit.

Example 2 — Size of Earth. Earth's real diameter is about 12,760 kilometers, which we'll approximate by rounding up to 13,000 kilometers. Dividing by 10 billion, Earth's scaled size is

$$\frac{13{,}000 \text{ km}}{10{,}000{,}000{,}000} = 0.0000013 \text{ km}$$

In this case, the small size of Earth means we are better off converting the answer to millimeters, remembering that there are 1,000 millimeters in a meter:

$$0.0000013 \text{ km} \times \frac{1{,}000 \text{ m}}{1 \text{ km}} \times \frac{1{,}000 \text{ mm}}{1 \text{ m}} = 1.3 \text{ mm}$$

Earth is a little over 1 millimeter (1/20 inch) in diameter on the Voyage scale, which is actually a bit smaller than the average ball point in a pen.

Example 3 — Earth-Sun distance. Recall that the Earth-Sun distance is about 150 million kilometers (1 AU). Dividing by 10 billion, this distance becomes

$$\frac{150{,}000{,}000 \text{ km}}{10{,}000{,}000{,}000} = 0.015 \text{ km}$$

This time it makes the most sense to convert to meters:

$$0.015 \text{ km} \times \frac{1{,}000 \text{ m}}{1 \text{ km}} = 15 \text{ m}$$

The Earth-Sun distance is about 15 meters (49 feet) on the Voyage scale.

TN30 For classroom teachers, I've posted an activity that you can use with students to explore the Voyage scale; see Section 1.2.3 of grade8science.com.

TN31 Here are a few related notes about human missions to Mars, since these are likely to come up in discussions with students. First, on the travel time: The lowest cost (minimum energy) way to get to Mars is on what is called a "Hohmann transfer orbit," which takes roughly nine months in each direction and would require staying on Mars for a little less than two years. The total roundtrip time would be close to three years. Shorter trips are possible if you use faster rockets. The six months in the example is faster than is currently (in 2025) possible but should be in reach of rockets that are under development (such as the SpaceX "Starship"). Future technologies, such as nuclear rockets, could reduce the travel time further. In that case, it might be possible to time the trip with orbital motions to reduce the roundtrip time to as little as about eight months.

Of course, there are many other challenges in sending humans to Mars. For example, while movies like *The Martian* (worth watching, and the book, by Andy Weir, is also worth reading) envision growing food locally on Mars, early missions will almost certainly require bringing enough supplies to keep the astronauts alive for their entire trip. There's also the issue of the fuel for the return flight, which would add so much weight to the trip that it would likely make the trip cost and energy prohibitive. For that reason, most proposals for human trips to Mars envision making the fuel for the return trip with ingredients available on Mars. That is likely to be possible — but you probably wouldn't want to count on it unless it has first been demonstrated by robotic spacecraft on Mars. Perhaps the most challenging issue for human trips to Mars will be radiation. Space is filled with dangerous radiation (mostly from the Sun). Here on Earth, we are protected by our atmosphere and Earth's magnetic field; the latter provides significant protection even in low-Earth orbits, such as that of the International Space Station. But a journey to Mars would mean months without that protection, which might well prove lethal unless we come up with much better ways to shield the interiors of spacecraft from radiation. The astronauts will also face mental challenges from the many months in a small spacecraft and physical challenges from the effects of prolonged weightlessness.

Despite all these challenges, it seems likely that humans will find a way to reach Mars successfully, perhaps even within the next decade or two. This leads to the question of whether we will actually "colonize" Mars. It's fairly easy to envision a permanent research station, with astronauts heading to Mars for two-year stints. But would anyone actually want to live there permanently? Personally, I'm doubtful. You'd never be able to go outside without a spacesuit, and the danger of radiation on the surface means you'd most likely be living underground. While it might sound adventurous to live on an entirely new world, I suspect that once people got there, they'd quickly discover that they'd rather head back home. (For further discussion of the challenge of Mars colonization, see *A City on Mars* by Kelly and Zach Weinersmith.)

TN32 Alpha Centauri is the nearest "star" in the night sky; in fact, its name means it is the brightest ("alpha") star in the constellation called Centaurus. But I put "star" in quotes because the Alpha Centauri system actually contains *three* stars: two that orbit each other fairly closely, called Alpha Centauri A and Alpha Centauri B, and a third that is somewhat closer to us (by about 0.1 light-year), called Proxima Centauri. This means that Proxima Centauri is technically the nearest star besides the Sun. However, Proxima Centauri is so much smaller and dimmer than our Sun that it cannot be seen with the naked eye, and the light of the "star" that we see as Alpha Centauri in the night sky comes almost entirely from Alpha Centauri A. So for practical purposes, it's usually easier to refer to the nearest star (besides the Sun) as simply being "Alpha Centauri." An added advantage of this choice is that Alpha Centauri A happens to be very similar in size and brightness (luminosity) to our

Sun, so on the Voyage scale you can visualize it as another grapefruit located about 4,300 kilometers (2,700 miles) from the grapefruit-size model Sun.

TN33 The math for calculating the distance of a light-year on the Voyage scale is particularly easy if you use powers of 10. Remember that a light-year is about 10 trillion kilometers, and 10 trillion is a 1 followed by 13 zeros, or 10^{13}. The Voyage model uses a scale of 1 to 10 billion, and 10 billion is a 1 followed by 10 zeros, or 10^{10}. We therefore find the scaled distance of 1 light-year by dividing the actual distance of 10^{13} km by the scale factor of 10^{10}. Remembering that dividing powers of 10 means subtracting exponents, we find

$$\frac{10^{13}\text{ km}}{10^{10}} = 10^{13-10}\text{ km} = 10^{3}\text{ km}$$

The result shows that a light-year becomes 10^3, or 1,000, kilometers on the Voyage scale.

TN34 Here's a little more detail about the stars that you can see by eye in the night sky. The number of stars visible to the naked eye depends both on your eyesight and on the darkness of the sky where you are observing. Catalogs and star charts generally describe how bright a star appears in the sky by its *magnitude*, with smaller or negative magnitudes representing brighter stars. For example, Sirius, which is the brightest star in our sky, has magnitude –1.46. Betelgeuse (the upper left shoulder of Orion) is the 10th brightest star with magnitude 0.50, and the "north star," Polaris, has magnitude 1.98 (making it the 47th brightest star in the sky).

The limit of human eye detection is generally considered to be about magnitude 6.5, and there are a total of about 9,000 stars that are at least this bright (that is, magnitude 6.5 or lower). Since only about half of these will be above the horizon at any one time, the best eyesight at the darkest sky sites will allow you to see about 4,500 stars.

You might at first guess that these 4,500 would be among the nearest of all stars, and while there's a bit of truth in that, the situation is actually a bit more complex. The reason is that the brightness of a star in our night sky depends both on its distance *and* on its intrinsic brightness, or *luminosity*. The vast majority of stars are intrinsically quite dim (compared to the Sun). For example, I've already noted that the nearest of all stars, Proxima Centauri, is so dim that it cannot be seen with the naked eye (see Teacher Note 32). At the other extreme, some stars are so intrinsically luminous that they can be seen even at fairly substantial distances. For example, the star Deneb, which is one of the three stars that make up what we call the "summer triangle" in the night sky, is the 19th brightest star in the sky despite being located more than 2,000 light-years away. On our scale where the galaxy fits on a football field, this puts Deneb about 2 yards (or meters) away, which means it is farther away than *millions* of other stars. It outshines these millions of nearer stars because it is unusually luminous (about 200,000 times as luminous as our Sun).

To summarize: Most people's arms extend outward from the center of their bodies by less than a yard (or meter), which means less than about 1% of the diameter of our football field galaxy (Figure 3.13). Since the real diameter of the galaxy is about 100,000 light-years, this means that if you spin around with your arms extended, you'll be tracing out a region with a radius representing less than about 1,000 light-years. While almost all the stars that are visible to the naked eye will be within this region, a few (like Deneb) will be a bit beyond it.

TN35 Radio waves are a form of light, and all forms of light travel through empty space at the same speed of light. More technically, "light" is what physicists describe as electromagnetic radiation, and all light is characterized by a wavelength (and frequency and energy). The visible light that our eyes can see has wavelengths ranging from about 400 to 700 nanometers. (A nanometer is a billionth of a meter.) But there are other forms of light with shorter and longer wavelengths. Radio waves are on the longest wavelength end, with infrared lying between radio waves and visible light. On the shorter wavelength side of visible light we find the kinds of light we call ultraviolet, x-rays, and gamma rays.

TN36 The actual number of stars in our galaxy is probably several hundred billion (and perhaps as many as 1 trillion). I've chosen to use the number 100 billion because it keeps things simpler. In case you are curious about the uncertainty in the number of stars: We know that the mass of stars in the disk of the galaxy is close to 100 billion times the mass of the Sun. However, we also know that most stars are less massive than the Sun. This is because while the Sun is "average" in the sense of being in the middle of the range of mass (and luminosity) for stars, smaller stars are more common, so the Sun is actually in the top 10% or so of stars by mass. Therefore, a mass of 100 billion Suns would mean several times that many stars.

TN37 The given quote is from the last page of my children's book *I, Humanity*, which uses a first-person narrative to tell the story of how we as a species have learned about our place in the universe. You can find more info about the book at bigkidscience.com/books/i-humanity/, and you can watch an astronaut reading it from the International Space Station in the video library of Story Time From Space (storytimefromspace.com).

T38 This particular image (Figure 4.1) from the Euclid space telescope shows galaxies ranging in distance to up to almost 11 billion light-years away — not quite as far as the most distant galaxies of Figure 1.1 from the James Webb Space Telescope (JWST), but still spanning much of the history of the universe. You can learn more about this image by searching its name, "Euclid Deep Field South," or the galaxy cluster around its center, which is called J041110.98-481939.3.

You might wonder why astronomers need the Euclid space telescope when they already have JWST and many ground-based telescopes, including the Rubin Observatory (see Teacher Note 8). The reason is because each of these telescopes has different key capabilities, which means we can get a much fuller understanding of the universe by using data from all of them together. For example, Euclid has a much wider field of view than JWST, which allows it to make a 3-dimensional map of galaxy distribution across a large portion of the sky — with the tradeoff that it cannot see quite as deeply into the universe. Euclid's primary advantage over Rubin (and other ground-based telescopes) is its ability to observe both visible and near-infrared light — with the tradeoff of being a smaller telescope that therefore cannot see quite as faint objects.

TN39 More specifically, the best current estimates place the age of the universe at about 13.8 billion years, with an uncertainty of about 0.2 billion years (200 million years). So the value I've given of "about 14 billion years" is a rounded value that is easier to remember.

Note that there are two major ways in which astronomers estimate the age of the universe. The first is the one I've mentioned in the main narrative, which is by using the expansion rate to calculate how long it has taken for galaxies to reach their current average distances. The uncertainty with this method stems primarily from the uncertainty in measurements of the current expansion rate and the fact that we are still learning about how the expansion rate has changed with time.

The second method comes from the "cosmic microwave background," which is remnant radiation from the heat of the Big Bang (see Figure 4.5). Careful study of this radiation essentially provides astronomers with a picture of how the universe expanded in its very early history, and astronomers can then use models based on current understanding of physics to estimate an age for the universe.

Reassuringly, both methods find very similar ages. They are not an exact match, however, which explains why you may sometimes see debate in the media about how well scientists know the age of the universe. Keep in mind that while this debate is interesting, the argument is over relatively small discrepancies that don't affect the basic idea that the universe is "about" 14 billion years old. If you'd like to learn more about this debate, look up the "Hubble tension," which is the name given to the discrepancies between the two methods.

TN40 Upon learning that our universe is expanding, students often ask whether this also means that Earth and other small objects — including us — expand with time. The answer is no. Just as individual galaxies are not expanding because gravity keeps them intact (even while the universe as a whole expands), the same is true for everything within galaxies. More generally, you can think about it this way: The expansion of the universe is an expansion of space itself. However, if an individual object is held intact by some other force — such as gravity in the case of a galaxy (or star or planet) or the interatomic forces that hold our bodies together in the case of us — then the space within that object will not expand with time. Note that the raisin cake analogy (Figure 4.3) shows this fairly well, since the raisins themselves do not get larger even as the cake as a whole expands.

TN41 This brings up another topic that is often covered in the media: Astronomers are having a difficult time understanding the early stages of galaxy formation. In particular, images like the JWST deep field in Figure 1.1 suggest that at least some galaxies were already forming only a couple of hundred million years after the Big Bang, which seems too fast according to current understanding of how gravity should have clumped material together in the early universe. This mystery persists even when astronomers account for the fact that most of the gravity must have been due to dark matter (see page 24), which represents the majority of the mass in galaxies and the universe. Note, however, that although this mystery means there are likely some significant gaps in our understanding of the *details* of galaxy formation, it does not imply any major problems with our general understanding of the history of the universe.

TN42 Two types of measurement are needed to conclude that galaxies are moving away and that more distant ones are moving faster: (1) We need to know galaxy speeds and directions, and (2) we need to know galaxy distances.

The first measurement turns out to be relatively easy, thanks to something called the *Doppler effect*. You have probably experienced the Doppler effect with sound, which you can hear in the dramatic change from high to low pitch — a sort of "weeeeeeee–ooooooooooh" sound — as a car or train passes you by. The Doppler effect also occurs with light, causing the light from a moving object to be shifted to shorter wavelengths (a blueshift) if the object is moving toward us and to longer wavelengths (a redshift) if it is moving away from us. The amount of the wavelength shift tells us the object's speed. (Traffic police use this effect to catch speeders with radar or lidar.) All galaxies beyond the Local Group have a redshift, which tells us that they are all moving away from us, and the amount of redshift tells us their speeds.

The bigger challenge is measuring galaxy distances, which is what Hubble needed to do in order to recognize that the more distant ones are moving faster. Hubble used fairly crude estimation techniques that nevertheless were good enough to establish the basic idea. Today, astronomers have much more sophisticated distance measurement techniques that allow the rate of expansion to be pinned down with a fair amount of precision, though some small uncertainties still remain — which is why there is some uncertainty both in the precise expansion rate and in the age of the universe.

TN43 The theory of relativity has a reputation for being hard, but while a complete understanding of it does indeed require advanced mathematics, it is possible for anyone of about middle school age or above to understand the basic ideas. Moreover, besides helping you understand the expansion of the universe, these basic ideas lead to many fascinating ideas that you may have heard about, such as the famous equation $E = mc^2$, why we can't travel faster than the speed of light, and the exotic objects known as black holes. There are many excellent books and videos that can introduce you to the ideas of relativity; I've even written one myself, called *What Is Relativity?* (available from Columbia University Press, more info at bigkidscience.com/relativity).

TN44 In case you are wondering how scientists know the real times at which events marked for Earth on the cosmic calendar occurred, including the formation of Earth and the solar system: In most cases, ages are measured from careful study of rocks and fossils. The most precise ages come from radiometric dating, a technique that relies on the fact that radioactive materials can serve as highly accurate clocks. Today, these dates are known so well that we can say with confidence that our solar system began to form about 4.57 billion years ago, with an uncertainty of only about 0.01 billion years (10 million years). Because it took the planets a few tens of millions of years to form, we conclude that Earth formed by about 4.50 billion years ago. You can learn more about how we date rocks and fossils (and learn the age of Earth) in Section 5.1.3 of grade8science.com.

TN45 An abundance of evidence indicates that a major impact occurred at the time the dinosaurs went extinct, though there is some debate over whether this impact was the sole cause or one of several causes for their extinction. Learn more about this evidence in Section 5.4.2 of grade8science.com.

TN46 Studies show that an actual blink of the eye generally takes between 0.1 and 0.4 second, which represents between about 40 and 160 years on the cosmic calendar (since 1 second represents a little more than 400 years). So it's quite accurate to say that a human lifetime is a blink of the eye on the cosmic time scale.

TN47 As this is the final Building a Cosmic Perspective question in this book, it provides a great opportunity for you (or your students) to think again about what a "cosmic perspective" actually means. I obviously hope that this book has provided you with insights on that, but in addition: (1) If you haven't already watched or read Carl Sagan's Pale Blue Dot speech, which I recommended earlier (see bigkidscience.com/Sagan), I hope you will do so now. (2) Neil deGrasse Tyson has a wonderful essay posted about his view of the cosmic perspective, which you can find at neildegrassetyson.com/essays/2007-04-the-cosmic-perspective/ (this essay also appears as a Foreword to my astronomy textbook *The Cosmic Perspective*).

To Learn More

I hope that this book has given you a general sense of the scale of the universe, but I also hope that you will be inspired to want to learn more. There are of course many great resources out there, but if you like this book, you may be interested in some of the other books and media that I've created for space, science, math, and education. Please check them out by visiting my publishing web page, where the various tabs will tell you about books, school programs, the Totality app, and other activities and resources:

www.BigKidScience.com

Acknowledgments

This book carries only my name as author, but it is the result of collaborative work with many others. I first began developing my approach to teaching about the scale of the universe in the late 1970s, strongly influenced by reading and watching the works of the great astronomer and communicator Carl Sagan (1934–1996). I then had the opportunity to develop these ideas through my early teaching career at Sunset View Elementary School in San Diego (under the mentorship of an incredible teacher, Anne Earlywine) and at a summer school that I created and taught for elementary and middle school students in a classroom that was graciously provided by Point Loma Nazarene College. While I no longer remember all the names, the many students and parents I worked with provided invaluable feedback about what worked and what didn't when it came to teaching about scale (and other topics).

The next step in developing my approach to teaching scale came with the development of the Colorado Scale Model Solar System, which was first installed at the University of Colorado (Boulder) in 1987. Although I started this project, much of the work was done by faculty advisor Tom Ayres (who was also my thesis advisor) and three undergraduate students at the time: Ron Bass, Matt Carter, and Ken Center. This project led directly to the Voyage scale model solar system that first opened in 2001 on the National Mall in Washington, DC. Many people were involved in that project, but I'd especially like to thank my friend and colleague Jeff Goldstein, who led the development and continues to run the Voyage National Program (voyagesolarsystem.org) today. Jeff and I have had dozens of great hours of discussion about how best to get scale ideas across.

Those ideas about how to teach scale ultimately became part of the astronomy textbook, *The Cosmic Perspective*, that I still write with co-authors Megan Donahue, Nick Schneider, and Mark Voit. The three of them helped me further refine both the general approach and specific words we use for many scale concepts. Indeed, much of the material in this book is adapted from *The Cosmic Perspective*. I thank our publisher, Pearson Education, for providing permission for me to adapt numerous art pieces from the textbook for this book.

The production of this book was made possible by the amazing designer Mark Ong. Mark has done the design work on all Big Kid Science books to date and has also worked on the design of my textbooks from both Pearson and Princeton University Press. Numerous people provided editorial input, but I especially thank Sally Lifland for her careful editorial review and Mark Voit, Richard Gelderman, and Jimmy Negus for their careful reviews of scientific accuracy. Other reviewers included Fran Bagenal, Julie Danneberg, Terry Bramschreiber, Laura Israelsen, Kent Schnacke, Michelle Pearson, William Meadows, Joshua Roth, Amir Said, Mark Levy, Helen Zentner, and Joanne Rhodes.

Finally, I thank my wife, Lisa, and my adult children, Grant and Brooke, for their ongoing support, inspiration, and insights. Their kindness and thoughtfulness embody everything that can help us create a better future for all.

Glossary

Alpha Centauri The nearest "star" in our night sky; actually a three-star system located about 4.3 light-years away. The nearest of the three stars is known as Proxima Centauri. (The others are known as Alpha Centauri A and B.)

Andromeda Galaxy (M31; the Great Galaxy in Andromeda) The nearest large spiral galaxy to our own Milky Way, located about 2.5 million light-years away.

Artemis program NASA's program for returning humans to the Moon.

asteroid A relatively small and rocky object that orbits a star.

astronomical unit (AU) The average distance between Earth and the Sun, which is about 150 million kilometers (93 million miles).

Big Bang The name given to the event thought to mark the birth of the universe; equivalently, the event marking the beginning of the expansion of the universe.

Big Bang theory The scientific theory of the universe's earliest moments, stating that all the matter in our observable universe came into being at a single moment in time as an extremely hot, dense mixture of subatomic particles and radiation.

center of the universe syndrome (COTUS) In the context of this book, a condition exhibited by people who behave as though they believe the universe revolves around them.

comet A relatively small and ice-rich object that orbits a star.

cosmic microwave background Radiation that comes from all directions in space and that matches the characteristics expected for radiation left over from the Big Bang.

cosmic web The web-like pattern with which galaxies (and galaxy clusters and superclusters) appear to be arranged when we look at very large regions of the universe.

dark energy The name given to the energy or force presumed to be responsible for the observed acceleration of the expansion of the universe. Its nature is a great mystery.

dark matter The name given to matter that we infer to exist because of its gravitational influence but from which no light has been detected. Its nature is a great mystery.

dwarf planet An object that orbits a star and is large enough for its gravity to make it round but not large enough to count as a "full" planet. (Note: See Teacher Note 10 for more detail on the distinction between planets and dwarf planets.)

Earth-ish planet A planet that is similar enough in size and orbit to Earth to make us think that it *might* be Earth-like (with oceans and an atmosphere that could support life). However, our current telescope technology does not allow us to determine whether such planets are truly Earth-like.

Earth-like planet A planet with a nature similar to that of Earth, generally meaning that it has oceans and an atmosphere that could in principle support Earth-like life.

exoplanet A planet orbiting a star other than our Sun.

expansion of the universe The observed fact that, on average, distances between galaxies are increasing with time.

fusion (nuclear) The process in which two or more small atomic nuclei are fused together to make a more massive nucleus.

galaxy A great island of stars in space, containing from a few million to a trillion or more stars, all held together by gravity and orbiting a common center.

galaxy cluster A relatively large group of galaxies held together by gravity, typically with hundreds or even thousands of galaxies.

galaxy group A group of galaxies held together by gravity. The term *group* is usually used when there are up to a few dozen galaxies; larger groups are called galaxy *clusters*.

gravity A force that attracts masses together. In most cases, more massive objects have stronger gravity. More specifically, the strength of gravity can be calculated from the universal law of gravitation, which states that the strength of gravity increases with mass but decreases with size or distance from the center of an object.

Great Red Spot A large, reddish-colored storm on Jupiter.

ice (in space science) Any substance that freezes solid at low temperatures. The most common ices are frozen versions of water (H_2O), methane (CH_4), ammonia (NH_3), and carbon dioxide (CO_2); the latter is sometimes called "dry ice."

interstellar Literally, "between the stars."

interstellar cloud A cloud of gas found between stars in space.

Kuiper belt (*Kuiper* rhymes with *piper*) The donut-shaped region of our solar system beyond the orbit of Neptune that contains vast numbers of comets having orbits that lie fairly close to the plane of planetary orbits and traveling around the Sun in the same direction as the planets.

Laniakea Another name for our local supercluster.

light-year (ly) The distance that light can travel in 1 year, which is about 10 trillion kilometers (6 trillion miles).

Local Group The group of a few dozen galaxies to which our Milky Way belongs. The Andromeda Galaxy is also part of the Local Group.

Local Supercluster (Laniakea) The supercluster of galaxies to which the Local Group belongs.

Milky Way The name used both for our galaxy and for the band of light we see in the night sky when we look into the plane of our galaxy.

moon (or satellite) An object that orbits a planet, though the term is also used for objects orbiting dwarf planets and for other smaller objects that orbit larger ones.

multiverse A hypothetical set of universes, in which our universe would be just one of many. Note that while this speculative idea is popular even among many scientists, we do not currently have any clear evidence that a multiverse actually exists.

observable universe The portion of the entire universe that, at least in principle, can be seen from Earth.

Oort cloud (*Oort* rhymes with *court*) A huge, spherical region centered on the Sun, extending perhaps halfway to the nearest stars, in which trillions of comets orbit the Sun with random inclinations, orbital directions, and eccentricities.

planet A moderately large object (currently taken to mean at least about a third the diameter of Earth) that orbits a star and shines primarily by reflecting light from its star.

satellite A term often used as a synonym for *moon*, though some scientists prefer to reserve the term *satellite* for human-made objects in space.

solar system The Sun and all the material that orbits it, including the planets. The term technically refers only to our own star system (because *solar* means "of the Sun"), but it is often applied to other star systems as well.

star A large, glowing ball of gas that generates heat and light through nuclear fusion in its core. Our Sun is a star.

star stuff In the context of stellar lives, material manufactured by stars that then becomes incorporated into future generations of stars and star systems. Hence we can say that we are made of "star stuff."

star system A star (or sometimes more than one star) and any planets and other materials that orbit it.

supercluster (of galaxies) The largest known structures in the universe, consisting of many clusters of galaxies, groups of galaxies, and individual galaxies.

transit In the context of astronomy, the phenomenon in which a planet passes directly across the face of its star as seen from Earth. A transit can only occur (as seen from Earth) if the planet's orbit happens to have no inclination to our line of sight to the star.

transit method An indirect method of detecting exoplanets by observing the dips that occur in a star's brightness each time an orbiting planet transits (passes in front of) the star, temporarily blocking some of the star's light.

universe (or cosmos) The sum total of all matter and energy; that is, all galaxies and everything within and between them.

Index